Fuzzy Management Methods

Series Editors

Andreas Meier
University of Fribourg, Switzerland

Witold Pedrycz
University of Alberta, Alberta, Canada

Edy Portmann
University of California at Berkeley, CA, USA

For further volumes:
http://www.springer.com/series/11223

Daniel Fasel

Fuzzy Data Warehousing for Performance Measurement

Concept and Implementation

Springer

Daniel Fasel
Scigility Inc.
Marly, Switzerland

ISSN 2196-4130 ISSN 2196-4149 (electronic)
ISBN 978-3-319-04225-1 ISBN 978-3-319-04226-8 (eBook)
DOI 10.1007/978-3-319-04226-8
Springer Cham Heidelberg New York Dordrecht London

Library of Congress Control Number: 2014933209

Printed on acid-free paper

Springer is part of Springer Science+Business Media (www.springer.com)

For my daughter Leola Mai Ly

Foreword

Data warehouses provide an important source of information for enterprises by homogenizing and concentrating business relevant data. In order to retrieve business critical information from the huge collection of data in the warehouse, data must be interpreted in the context of the information one wants to obtain from them. Data in the warehouse, however, are usually imprecise due to imprecisions in the collected data themselves or imprecisions coming from the way collected data is processed when integrated in the warehouse. There is another source of imprecision: the retrieval of information from data. As data are just facts, say the size of the customer data base, the mere fact does not say anything, for instance, about the competitiveness of the enterprise, and it may not be obvious which facts must be interpreted in which way to finally get such information.

On the other hand, dealing with imprecisions in data or in the way they are interpreted is not new. The concept of a weighted membership of an element in a set was already introduced in the mid-1960s by Lotfi A. Zadeh and Dieter Klaua. The resulting fuzzy sets can be used to "fuzzify" the classification of data using certain keywords to describe the information content of data. The size of the customer data base, for example, is then not a number any more, but it may be to 60 % *big* and to 40 % *mid-sized*.

This is exactly the observation Daniel Fasel makes in his thesis *Concept and Implementation of a Fuzzy Data Warehouse* that is published in this monograph and for which he received his PhD in 2012 from the Economics and Social Sciences Faculty of the University of Fribourg in Switzerland. He develops a concept of fuzziness in a data warehouse, demonstrates its applicability using the example data warehouse of a movie rental company, and finally develops a prototype implementation of his concept. With that, he provides us with a convincing solution for dealing with imprecisions in data warehouses.

Daniel Fasel's monograph is structured as follows. After an introduction in Chap. 1, he provides in Chap. 2 all necessary background information on data warehouses. In particular, he presents the star and snowflake schema that he uses later on when developing his own approach. Chapter 2 also gives an overview of fuzzy logic to the extent needed in the remainder of the monograph. Chapter 3 is conceptually the main chapter of the monograph. There the concept of the fuzzy data warehouse is developed. Existing approaches are presented and the relation of the author's concept with existing research is discussed critically. Then the fuzzy data warehouse is developed in detail. It leads to the fuzzy data warehouse meta model, consisting of a classical data warehouse accompanied by a fuzzy data warehouse that integrates fuzzy classification into the classical warehouse model. Also included is a detailed section on how to fuzzify the data. This is absolutely crucial when realizing such a warehouse concretely. Then also the operations in a fuzzy data warehouse are discussed in detail, including the somewhat surprising operation that is the opposite to fuzzification: defuzzification. After having developed the conceptual warehouse model completely, by using the example of a data warehouse for a movie rental company, the sometimes a little abstract concepts developed in Chap. 3 are explained very comprehensively in Chap. 4. Finally, the entire approach is implemented using Java and XML in Chap. 5. I do find the implementation chapter absolutely crucial as it demonstrates that all abstract concepts (even Chap. 4 is conceptual in that it shows the applicability of the approach only to a single example scenario) can be realized in practice. This monograph concludes with a chapter critically evaluating the presented approach.

Having reviewed Daniel Fasel's thesis, I must definitely state that its content that is also presented in this monograph is an excellent piece of applied research in the field of data warehouses. In fact, the development of the fuzzy data warehouse meta model represents even a fundamentally new concept. The presented new approach is particularly convincing in that it does not stop with developing the general framework and evaluating it using the example warehouse of a movie rental company, but indeed delivers an excellent prototype implementation. Going that far guarantees that the fuzzy data warehouse is not just a nice research idea, but a viable approach for developing a new type of data warehouse in practice. In addition to its very interesting content, this monograph is also written in a very comprehensible way.

Without hesitation I can recommend this monograph to its readers (who will come in particular from the field of data warehouses) as I strongly believe that they will gain new insights in to data warehouses by reading about the presented novel concept of the fuzzy data warehouse.

Fribourg, Switzerland
4 September 2013

Prof. Dr. Ulrich Ultes-Nitsche

Acknowledgment

I would like to thank Prof. Dr. Andreas Meier. Without his support and encouragement I would never have started this thesis, much less I would have finished it. During the PhD study, he always provided invaluable input for the thesis with his academic advices. His enthusiasm has been a great motivation during this time.

I also wish to thank Prof. Dr. Ulrich Ultes-Nitsche for his constructive feedback.

I would also like to express my deepest gratitude to Khurram Shahzad. Without his inputs, this thesis would never be in the shape as it is now. The time I spent for research with Khurram has been the most productive moments during my PhD study. In Khurram, I found a very very good friend and I am more than glad that I met Khurram through our common research interests.

I would also like to express my gratitude to my research colleagues in the Information System Research Group and in the Department of Informatics of the University of Fribourg.

Many thanks go to Kristen Curtis and my brother Thomas, who read the thesis and helped me a lot improving the language.

Last, I would like to thank my wife Stéphanie and my daughter Leola. Without your patience and support I would never have had the energy to write this thesis. You are the essence in my life and the reason for writing this thesis, for doing what I am doing. With every single word in this thesis I would like to tell you: I love you!

Contents

List of Abbreviations

AJAX	Asynchronous JavaScript and XML
API	Application programming interface
AVG	Average function
BFN	Joint Master Program of the Universities of Fribourg, Bern and Neuchatel
CHF	Swiss Franks currency
CMA	Class membership attribute
DiUF	Department of Informatics of the University of Fribourg
DWH	Data warehouse
ETL	Extraction, transformation and load
FC	Fuzzy concept
FCONCEPT	Fuzzy concept
FCT	Fuzzy classification table
FDW	Fuzzy data warehouse
FK_	Foreign key prefix
FMT	Fuzzy membership table
HK	High knee point
JAXB	Java Architecture for XML Binding
JDBC	Java database connector
LK	Low knee point
MAX	Maximization function
MD	Membership degree
MDA	Membership degree attribute
MDX	Multidimensional expression language
MEMFUNC	Membership function
MIN	Minimization function
MOLAP	Multidimensional online analytical processing
NAT	Faculty of Mathematical and Natural Sciences of the University of Fribourg
OLAP	Online analytical processing
PK_	Primary key prefix

PROD	Mathematical product
ROLAP	Relational online analytical processing
SES	Faculty of Social and Economic Sciences of the University of Fribourg
SQL	Structured Query Language
SUM	Summation function
TA	Target attribute
XML	Extensible Markup Language
XSD	XML Schema Definition

List of Figures

List of Tables

Listings

Chapter 1
Introduction

The introduction chapter provides an overview about the motivation of this thesis in Sect. 1.1. It summarizes how fuzzy logic in combination with data warehouses might be useful. In Sect. 1.2, the research methodology of the thesis and fundamental question this thesis aims to answer are depicted. An overview of the chapter structure is then given in Sect. 1.3. Finally, the research papers that have been published within the study of this dissertation are summarized in Sect. 1.4.

1.1 Motivation

Nowadays, information is an important good for enterprises. Information collected from the business environment is often the basis for decision making. In order to gather, store and process this information, various information systems are used. The enterprise information system map shows often numerous, heterogenous and complex information system constellations. Often, for operational use, relational database systems are used and for analytical purposes a data warehouse is used.

The data warehouse systems receive its data for analysis from the surrounding operational systems. Because of the complexity and the number of operational systems, the data received by the data warehouse is heterogenous. The data has to be homogenized in a first step before the data warehouse can further process it. After processing and loading the data into the data warehouse, the enterprise receives a centralized view of their business relevant information. These views in data warehouses provide different perspectives on business critical indicators, further called facts. The perspectives can be adapted according the need of the analysis. This adoption is realized by designing navigation paths, further called dimensions, in the data warehouse. Dimensions allow examining facts in different granularities and represents time, topologies and other business interesting perspectives. This analytical view on data finally enables the enterprise to have a more global sight

D. Fasel, *Fuzzy Data Warehousing for Performance Measurement*, Fuzzy Management Methods, DOI 10.1007/978-3-319-04226-8_1,

on its business environment than operational systems can provide. Therefore, data warehouses are often used as systems for decision making.

Besides positive aspects of centralized processing of business information such as decision making support, difficulties occur in maintaining and analyzing data warehouses. The amount of data that has to be processed in a data warehouse increases every day and turns into challenging tasks for administration and analysis. Next to the problem of high quantity, data from operational systems are often incomplete, vague or uncertain. This quality issue can not be completely eliminated in the preprocessing stage of the data. Consequently, a certain amount of vagueness directly impacts the analysis and decision making that is based on the information of a data warehouse.

In addition to the qualitative issues of the data entered in the data warehouse, the interpretation of the analysis can vary depending on the background of the user. A fact that provides a number of sales might be interpreted differently by a sales representative or by a financial controller. This various interpretation across endusers of reports makes it difficult to fulfill the analytical needs of an enterprise with a preferably small amount of reports from the data warehouse. Furthermore, even if endusers interpret a fact similarly, decision making is often vague and the consideration of a fact is approximative. The fact is interpreted in linguistic context such as "less sales" or "higher costs" which leads to a linguistic approximation of the fact. Standard analysis in data warehouses can not deal with this vagueness in language and therefore limits the interpretation and classification of data.

The theory of fuzzy sets and linguistic variables proposed by Zadeh in [Zad75a, Zad65] can deal with vagueness, uncertainty and imprecision. In contrast to probabilistic systems, it is also optimized for interpreting the imprecision in human language and reasoning. Consequently, the application of fuzzy logic on data warehouse technologies might improve the analysis of data and might lead to better decision making.

In this thesis, a fuzzy data warehouse modeling approach is presented, which allows integration of fuzzy logic as meta tables without affecting the core of a classical data warehouse. The essence of the approach is that meta table structure is added to a data warehouse for classification, which enables integration of fuzzy logic in dimensions and facts, while preserving the time-invariability of the data warehouse. Also in this thesis, a method is presented which includes guidelines that can be used to convert a classical warehouse into a fuzzy data warehouse. The key benefit of integrating fuzzy logic in meta tables is that it allows analysis of data in both sharp and fuzzy manners. In addition to that, the aggregation of fuzzy logic concepts and typical operations of a classical data warehouse are discussed for the fuzzy data warehouse. The use of the proposed approach is demonstrated through a case study of a movie rental company, a prototype implementation of the fuzzy data warehouse and the benefits of integrating fuzzy logic concepts in performance measurement are illustrated.

1.2 Research Methodology

In order to address the problems stated in the motivation, a design science research approach [HMPR04, MS95] has been chosen. The approach aims to link fuzzy logic and data warehouse technology together. Hevner provides a guideline in [HMPR04] for design science research. The next paragraphs illustrate how this guideline is applied in this thesis.

1.2.1 Problem Relevance and Research Rigor

In the first step, the problem of integrating fuzzy logic in a data warehouse has been analyzed and the following fundamental research questions have been formulated:

- How can facts in data warehouses be handled fuzzily?
- How can dimensions in data warehouses be handled fuzzily?
- How can data warehouse operations like roll-up, drill-down, slice and dice be applied to fuzzy data in a data warehouse?
- How can fuzzy data be aggregated?
- Is it possible to propose generic modeling methods for fuzzy data warehouses?
- How is it possible to integrate a fuzzy data warehouse into an existing information system infrastructure in a company in the easiest manner?
- What kind of technologies can be used for implementing a fuzzy data warehouse?
- How does a fuzzy data warehouse behave in comparison with a crisp data warehouse?

The conceptual design part was first started by analyzing existing literature. How the proposed fuzzy concepts can provide added values and what are the problems of the existing approaches has been analyzed. Based on existing literature and fundamental concepts of data warehousing and fuzzy theory, a new fuzzy data warehouse concept has then been developed.

1.2.2 Design as a Search Process and Design as an Artifact

Based on the analysis of the existing attempts to incorporate fuzzy logic with data warehouse technology, a summary of potential critics and problems has been done. In the next step, a meta table model has been developed which overcomes the major criticisms and still seems to be able to cover all of the applications presented in literature. The meta model has then been integrated into a fuzzy data warehouse meta model that describes the incorporation of crisp data warehouse technologies and fuzzy logic in the form of a fuzzy concept meta table structure.

Furthermore, a method has been developed that explains how this fuzzy data warehouse meta model can be applied to existing data warehouses. This proves that it is possible to integrate the meta table structure into existing data warehouses with a reasonable amount of effort.

1.2.3 Design Evaluation

In order to evaluate the fuzzy data warehouse concept, an experimental and descriptive approach has been chosen. Using a virtual movie rental company as an example, it is described how a classical data warehouse can be modeled into a fuzzy data warehouse using the proposed modeling method. Further, it is shown how the application of fuzzy concepts in analysis can improve the information quality of reports.

Based on the application, a prototype implementation is realized that proves the feasibility of a fuzzy data warehouse. The prototype is based on existing technologies that have already been used for developing data warehouse systems. Consequently, it can be derived from the prototype implementation that the application of a fuzzy data warehouse in existing data warehouse systems is possible.

1.2.4 Research Contribution and Research Communication

The contribution of the fuzzy data warehouse concept, the application and the prototype is discussed in detail in the evaluation of this thesis. First, how the proposed fuzzy data warehouse can solve the stated critics of the existing applications is described. In the second step, potential limits of the concepts are shown.

The above listed research questions are discussed in greater detail in the conclusion. The conclusion attempts to answer each research question and shows in which part of the thesis the question has been discussed.

The research on fuzzy data warehouse during this dissertation study has been communicated several times in the form of research papers for conferences and a book chapter. With the publication of partial results, feedback and criticisms from the research community could be obtained. This feedback and criticisms have then been used to improve the fuzzy data warehouse concept.

1.3 Chapter Overview

The subsequent chapters are grouped into four parts. Each part covers a major subject area of this thesis. The next paragraphs discusses the underlying chapters for each part and shortly describes the subject of the chapters.

1.3.1 Part I: Concept

The concept part discusses the development of the fuzzy data warehouse concept. The included chapters are composed as follows:

Chapter 2 includes a discussion about fundamental concepts like data warehouse theory and fuzzy logic. Section 2.1 depicts how data warehouse systems are built and structured. Then, the aggregation of facts is analyzed more closely. Based on the multidimensional cube definitions, the classical data warehouse operations are finally defined. The Sect. 2.2 introduces the fuzzy logic. The fuzzy set theory including some of the fundamental fuzzy set operations is depicted. In order to illustrate the usage of fuzzy logic, a few application domains for fuzzy logic are discussed.

Chapter 3 develops the fuzzy data warehouse concept and can be seen as the core chapter of this thesis. In order to develop the concept, existing approaches in literature are first analyzed in Sect. 3.1. Based on the identified problems, a new fuzzy data warehouse model is developed in Sect. 3.2. In order to facilitate the implementation of fuzzy data warehouses in existing infrastructure, Sect. 3.3 provides a method on how to model a fuzzy data warehouse on top of an existing classical data warehouse. Section 3.4 then discusses the characteristics of fuzzy concepts in the fuzzy data warehouse. It is specifically focussed on the aggregation behavior of fuzzy concepts. Finally, in Sect. 3.5 the classical operations are redefined and two new operations are defined for the fuzzy data warehouse to support crisp and fuzzy data.

1.3.2 Part II: Application

In the application part, the movie rental company example is developed. Based on this example, it is then discussed how a fuzzy data warehouse can be modeled and how analysis with fuzzy concepts improves the information quality. This part is composed of Chap. 4 which is structured as follows:

In the first Sect. 4.1, the movie rental company and its classical, crisp data warehouse is presented. The second Sect. 4.2 discusses the creation of fuzzy concepts for each dimension and fact. The development of the fuzzy concepts in this chapter is based on the modeling method presented prior in Sect. 3.3. After creating a fuzzy data warehouse from the classical data warehouse, Sect. 4.3 presents how the fuzzy data warehouse can be queried. Therefore, multiple analyses are created integrating fuzzy concepts and the benefit of the fuzzy concepts is depicted.

1.3.3 Part III: Implementation

After the discussion of the application, it is shown in this part how the fuzzy data warehouse can be implemented. A prototype fuzzy data warehouse application is created based on a three tier architecture as described by Eckerson [Eck95]. Similar to Part II, Part III only contains one chapter. The sections of Chap. 5 are as follows:

Section 5.1 first gives an overview of the architectural approach. In the next Sect. 5.2, the database tier is discussed. It is shown on which technology the databases tier is built, how stored procedures and triggers can be used for implementing the fuzzy concepts. The business tier is depicted in Sect. 5.3. Additional to the data in the database, some meta information of the fuzzy data warehouse has to be managed in the business tier. How the meta information can be modeled and maintained using XML schema definitions and XML documents is subsequently discussed. Further, the business tier has to translate the inputs from the visualization tier and pass them to the database tier. Therefore, two engines are illustrated that provide these functionalities. The last Sect. 5.4 depicts the visualization tier. This tier provides the user interface of the fuzzy data warehouse. How the user can navigate the fuzzy data warehouse and administrate the fuzzy concepts is then explained.

1.3.4 Part IV: Evaluation and Conclusion

The last part contains the evaluation of the fuzzy data warehouse and a conclusion. Therefore, Part IV includes Chap. 6 which is composed of the following sections:

Section 6.1 conducts the evaluation of the concept, the application and the implementation. It is divided into three sections. The first Sect. 6.1.1 contains the evaluation of the conceptual part. This section is further divided into a subsection depicting the benefits of the concept and another subsection looking at the limitations of the concept. The Sect. 6.1.2 is similarly structured and outlines the benefits and limitations of the application and the implementation. The last Sect. 6.1.3 discusses an outlook of the proposed fuzzy data warehouse.

Section 6.2 concludes this thesis. It first summarizes the thesis and points to the most important features of the concept, application and prototype. Then it briefly summarizes the evaluation and finally discusses how the defined research questions in Sect. 1.2 have been addressed in the thesis.

1.4 Publications

In this section the publications that have been completed during the dissertation study are outlined. Each publication is summarized and it is explained in which part of the thesis the content of the publication is incorporated.

1.4.1 *A Fuzzy Data Warehouse Approach for the Customer Performance Measurement for a Hearing Instrument Manufacturing Company*

This paper [Fas09] was presented at the Fuzzy Systems and Knowledge Discovery conference in China in 2009. It discusses a first approach of a fuzzy data warehouse for a hearing manufacturing company. The novelty of this approach is the implementation of fuzzy concept using meta tables. Furthermore, it is discussed how the concepts on facts can be aggregated using an arithmetic mean operation.

The meta table structure has been refined and finally integrated in the Sect. 3.2 discussing the fuzzy data warehouse concept. The aggregation of fuzzy concept has been integrated in Sect. 3.4.

1.4.2 *A Fuzzy Data Warehouse Approach for Web Analytics*

This paper [FZ09] was written together with Darius Zumstein for the conference Visioning and Engineering the Knowledge Society in Greece 2009. It illustrates the application of a fuzzy data warehouse and its meta tables for web analytics [Ass11, Has08, Kau07]. This paper shows a first feasibility study for the fuzzy data warehouse concepts.

Its application domain and conclusion has been integrated in the evaluation in Sect. 6.1.1. There, it illustrates the usefulness of fuzzy concepts in a fuzzy data warehouse compared to classical crisp data warehouses.

1.4.3 *A Data Warehouse Model for Integrating Fuzzy Concepts in Meta Table Structures*

This paper [FS10] was written together with Khurram Shahzad and has been presented in the 17th IEEE International Conference and Workshops on the Engineering of Computer-Based Systems in Oxford, UK 2010. This paper refines the meta table approach for a fuzzy data warehouse and provides the foundation of the fuzzy data warehouse meta model and formal meta table definitions in Sect. 3.2. Further, it defines guidelines that propose how to implement variants of fuzzy concepts within the meta table structure. The fuzzy data warehouse concept part is derived from the outcome of this paper.

The findings of this paper are integrated in the following Sects. 3.2.1–3.2.4.

1.4.4 Fuzzy Data Warehouse for Performance Analysis

This book chapter [FS12] was written with Khurram Shahzad and is published in the book Fuzzy Methods for Customer Relationship Management and Marketing: Applications and Classifications, 2012. It summarizes the modeling approach of a fuzzy data warehouse discussed in [FS10] and exemplifies the fuzzy data warehouse on an application for a movie rental company. A method for modeling a fuzzy data warehouse has been proposed and the movie rental company fuzzy data warehouse has been modeled. Furthermore, it depicts characteristics of fuzzy concepts in a fuzzy data warehouse and shows how operation of the fuzzy data warehouse can be defined.

The application of the movie rental company is refined in the Chap. 4. The method for modeling a fuzzy data warehouse is discussed in the Sect. 3.3. The characteristics section of this book chapter is the basis of the discussion about characteristics of fuzzy concepts in Sect. 3.4. Finally, the presented operations have been further refined and are discussed in Sect. 3.5.

1.4.5 Bachelor Thesis of Stefan Nüesch and Christoph Rathgeb

Stefan Nüesch developed in his bachelor thesis [Nüe11] an extensible OLAP cube in Java. Based on the data warehouse presented in Chap. 4 and the XML schema definition presented in Sect. 5.3, the OLAP cube provides simple navigation of the data through a web front end. Christoph Rathgeb extended in his bachelor thesis [Rat11] the OLAP cube in order to administrate and navigate with fuzzy concepts according Chap. 3.

Both bachelor thesis are reused in the Sect. 5.3.3 and in the Sect. 5.4 and are an essential part of the prototype implementation.

Part I
Concept

Chapter 2
Fundamental Concepts

2.1 Data Warehouse Concepts

Data warehouse was first discussed by Devlin and Murphy in 1988 [DM88]. They described a read-only database for integration of historical operation data and propose tools for user interaction with this database for decision support and analysis. However, Inmon's definition has received the most attention over the years. According to Inmon [Inm05], "a data warehouse is a subject oriented, non volatile, integrated and time variant collection of data in favor of decision making". One or more key business metrics, often extracted from different data sources, are **integrated** in the data warehouse and define the **subject** for analysis. The granularity of business metrics limits later the analysis capabilities of the data warehouse. Therefore, the level of granularity defines the information quality of the data warehouse.

In general, a data warehouse does not necessarily provide functionalities to modify or delete data once it is stored in the repository. Neither possess data in a data warehouse properties such as data in temporal databases. Whereas approaches for combining data warehouse and properties of temporal databases like valid-time have been proposed as data warehouse variants in [CS99, Ede01, MV00]. Due to the shortcomings of manipulation, deletion functionalities and the arrangement of data on a timeline, the data warehouse can be characterized as **non volatile** and **time variant**. As consequence, data size is steadily growing and a more sophisticated data management in the data warehouse is required than in transactional databases. Archiving and distribution capabilities of the data repository imply more overhead of meta data in a data warehouse [MT02]. Inmon [ISN08] identified difficulties and provides solutions to data and meta data handling of big distributed data warehouse systems.

Introduced in the mid 1990s, data warehouse has been widely adapted in several sectors and successful implementations have been reported in healthcare [Ion08], agriculture [Sch10], geo – information systems [MZ04], banking sector [HKYC04]

D. Fasel, *Fuzzy Data Warehousing for Performance Measurement*, Fuzzy Management Methods, DOI 10.1007/978-3-319-04226-8_2,

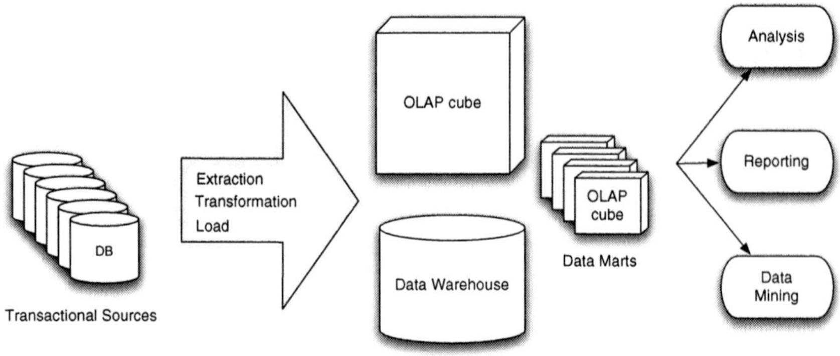

Fig. 2.1 Architecture of data warehouse (Adapted from [CD97])

and business environments [KR02] in order to enhance the analytical capabilities of enterprises.

Figure 2.1 shows the architecture of a classical data warehouse. Transactional sources act as a data source for data warehouse, however the data stored in transactional systems is not optimized for analysis and decision making [KRT$^+$08]. Therefore, data from transactional sources is extracted, transformed into a form that is compatible with warehouse after removing inconsistencies and loaded in a data warehouse for further usage [KC04]. As shown in Fig. 2.1, data warehouse can further be used for analysis using analysis tools like Online Analytical Processing (OLAP) and data mining to identify patterns for forecasting.

In contrast to classical databases in which entities and their relationships are modeled [Che76], in data warehouse, the values (key business metrics) that are critical for a business and the different aspects that can be used to analyze are modeled in a multidimensional way [CCS93]. Critical values are generally numeric values and are called facts. Examples of facts are quantity of products sold, amount of profit earned. Aspects are the different perspectives that can be used for analysis of facts and are called dimensions.

Dimensions can be organized hierarchically to support multilevel analysis – analyses at different granularity levels. An example of a hierarchy for a location dimension can be country, state, area, city and place. Similarly, an example of a hierarchy for time dimension can be year, quarter, month, week, date. The two examples can be used for analyses of quantity of products sold (a fact) from location perspective at country, state, area, city, place levels and per time unit.

The logical structure of data warehouse called dimensional schema (sometimes referred to as multidimensional schema) can be model with at least three model types: star, snowflake and fact constellation schema according Kimball et al. [KRT$^+$08]. These model types differ based on the difference in quantity and types of dimensions and facts. Out of the given types, star schema and snowflake schema are the most discussed ones. A schema is called star schema, if the dimensions surrounding a fact table are unnormalized, whereas a schema is called snowflake if the dimensions surrounding a fact table are normalized.

In contrast to Kimball, Inmon considers multidimensional data models as data marts [Inm05] or specialist data warehouse system [ISN08]. According to Inmon, multidimensional implementation of a data warehouse is a special field of application and limits usage and flexibility of the data warehouse. Multidimensional data warehouses are improved for using OLAP techniques. The term Online Analytical Processing was first characterized by E. F. Codd in 1993 [CCS93] and the OLAP Council [OLA10] defines it as "category of software technology that enables analysts, managers and executives to gain insight into data through fast, consistent, interactive access to a wide variety of possible views of information that have been transformed from raw data to reflect the real dimensionality of the enterprise as understood by the user." The data warehouse is the data repository and functionalities as the typical data warehouse operations roll-up, drill-down, slice and dice are defined by the OLAP engine. Despite of Inmons perception of a data warehouse system, one of the most common and most successful implementation of data warehouses is Kimball's multidimensional model in combination with OLAP techniques. Therefore, the term data warehouse is generally used as a synonym for the combination of a multidimensional data warehouse storage engine (OLAP server [OLA10]) and a OLAP engine as interface for accessing and analyzing the data. Further in this thesis, data warehouse will be considered as combination of OLAP server and engine.

The following Sects. 2.1.1 and 2.1.2 closer examine definitions of the multidimensional modeling approach for data warehouses. Characteristics of consistent aggregations will be discussed in Sect. 2.1.3. Section 2.1.5 will provide formal definitions of the classical data warehouse operations. These definitions are then further used to define a fuzzy data warehouse model in Chap. 3.

2.1.1 Dimension

Lehner [Leh03] defines a dimension of a data warehouse as a partial ordered set of category attributes with a functional dependency on each other and a generic top element. Moreover, a category attribute exists, which defines the lowest level of hierarchy of the dimension. Formally defined:

Definition 1 (Dimension [Leh03]). A scheme of a dimension D is a set of partial ordered category attributes $(\{D_1, \ldots, D_n, Top_D\}; \rightarrow)$ where $\rightarrow$ describes the functional dependency and Top_D is a generic top level attribute in relation to $\rightarrow$ in a way that Top_D is functionally defined by all category attributes $(\forall i (1 \leq i \leq n) : D_i \rightarrow Top_D)$. Moreover, exactly one D_i exists that defines all other category attributes and represents therefore the lowest granularity $(\exists i (1 \leq i \leq n) \forall j (1 \leq j \leq n, i \neq j) : D_i \rightarrow D_j)$.

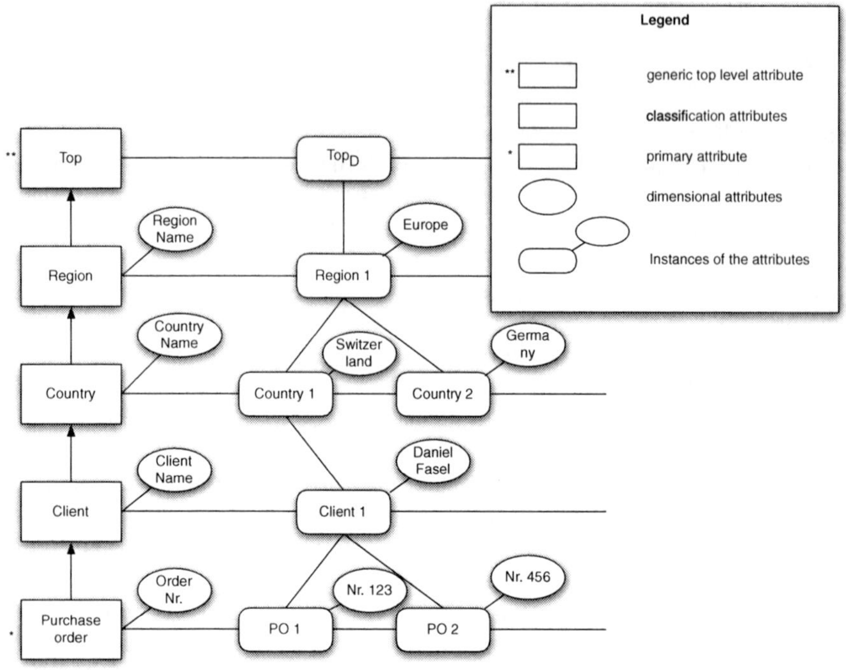

Fig. 2.2 Example dimension "Purchase Order Structure"

In a multidimensional model the dimension represents, following this definition, a hierarchical composition of different attribute elements. The category attributes can be sub-classified into the following groups:

- **Primary attribute**: This is the category attribute that represents the lowest granularity of a dimension. Only one of this type exists per dimension.
- **Top level attribute**: This is the generic top level attribute of a dimension. Only one top level attribute exists in a dimension.
- **Classification attribute**: This is the attribute defining the hierarchy of levels in the dimension, excluding the primary attribute and the top level attribute. Several attributes of this category may exist in a dimension and on one hierarchy level.
- **Dimensional attribute**: This attribute represents additional information, a property, on a specific dimension hierarchy level, respectively for a specific classification attribute. Every classification or primary attribute may contain several dimensional attributes.

The terminology of category attributes comes from statistical data based literature and is commonly used there [Sho82]. Category attributes have a finite set of disjoint discrete values which is also called category in the statistical terminology.

For example, Fig. 2.2 (following Lehner in [Leh03]) shows a possible dimension "Purchase order structure" with the category attributes: "Purchase order", "Client", "Country" and "Region". "Purchase order" is the lowest attribute and therefore considered as the primary attribute. "Client", "Country" and "Region" are classification attributes. The dimension ends with the top level attribute. Several dimensional attributes as "Order Nr." or "Client Name" are indicated in the figure as well. Additionally, the right hand tree structure shows instances of the different attributes of this dimension.

Dimensions play a fundamental role in multidimensional data warehouse systems. They are the framework for the navigation in the data warehouse. The actual information carrier are the facts. Dimensions are therefore used to aggregate, select and sort the facts and performance indicators.

2.1.2 Fact

A fact is defined by Lehner as a base performance measure that is in a certain granularity in the multidimensional data model [Leh03]. It is additionally characterized with a summation type that defines its basic behavior for the aggregation over dimensions and therefore the transition from one granularity level to the other. Formally:

Definition 2 (Fact [Leh03]). A fact F consists of a granularity G and a summation type *SumType*: $F = (G, SumType)$. The granularity $G = \{G_1, \ldots, G_n\}$ is a subset of all category attributes of all in the multidimensional data warehouse scheme existing dimensions with a dimension scheme $DS_1, \ldots, DS_n$ in a way that:

1. $\forall i (1 \leq i \leq n)\ \exists j (1 \leq j \leq n) : G_i \in DS_j$
2. $\forall i (1 \leq i \leq n)\ \forall j (1 \leq j \leq n) i \neq j : G_i \rightarrow G_j$

The summation type based on [Leh03] and [LS97] can either be a FLOW, STOCK or Value Per Unit type (a more detailed explanation can be found in Sect. 2.1.3).

Facts can be combined to form new performance indicators. As an example, the fact "Revenue" and the fact "Total production cost" can be combined to form the performance indicator "Profit". These performance indicators can then be handled the same way as facts concerning the aggregation over the dimensions. Lehner defines a performance indicator with a granularity, a computing method over a non empty set of facts and a summation type. Formally:

Definition 3 (Performance Indicator [Leh03]). A performance indicator K consists of a granularity G, a computing method $f()$ over a non empty set of facts F that exists in the multidimensional data warehouse scheme and a summation type $SumTyp : K = (G, f(F_1, \ldots, F_n), SumTyp)$.

The computing methods for a performance indicator can be classified in the following groups:

- Scalar function as $+, -, *, /, mod$
- Aggregation functions as $max(), min(), sum()$
- Ordering functions

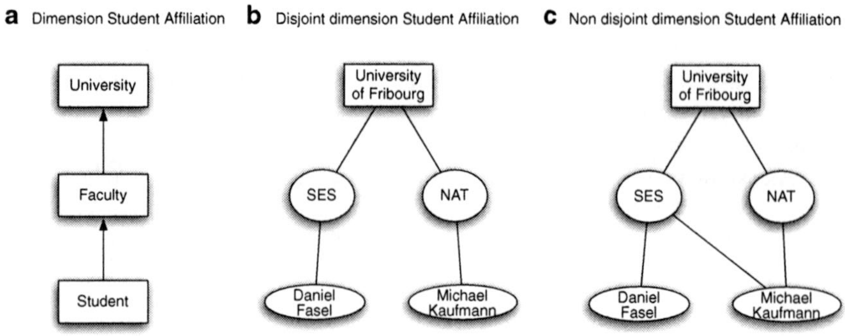

Fig. 2.3 Disjointness condition for dimension student affiliation

2.1.3 Summarizability

A data warehouse uses summarization functions for the aggregation of measures over the dimensions. Depending on dimensional relations, hierarchical structures can be complex and if the summarization of data is inconsistent, the result set of the corresponding operations is erroneous. Rafanelli and Shoshani [RS90] defined a framework called summarizability that describes characteristics for consistent aggregations. The framework is based on earlier discussions about operations in statistical databases (Shoshani [Sho82]). Based on Rafanelli and Shoshani [RS90] and Lenz and Shoshani [LS97] defined the necessary conditions to guarantee summarizability in OLAP cubes. Data summarization and dimension structures have to fulfill these conditions in order to provide consistent aggregations.

The first condition for summarizability states that aggregation of elements over a dimensional structure must always form disjoint subsets. Aggregating a fact over a dimension is disjoint if the category attributes form disjoint subsets over the fact instances. Considering the example of the affiliation of students to a faculty, the condition of disjointness is fulfilled if a student can only belong to one faculty at a time. If a student belongs to several faculties, it is not possible to consistently aggregate students over faculties anymore.

Figure 2.3a shows a dimension "Student Affiliation" with its hierarchy. The category attributes are "Student", "Faculty" and "University". In Fig. 2.3b a disjoint aggregation of the instances of the category attributes is shown. The category attribute "Faculty" builds disjoint clusters over the "Student" ("Daniel Fasel" only belongs to "SES" and "Michael Kaufmann" only belongs to "NAT") and so does "University" over "Faculty". With this dimension the calculation of the number of students at the university will provide the exact result of two students. Figure 2.3c shows a non disjoint aggregation. The student "Michael Kaufmann" belongs to more than one "Faculty" at the same time. Therefore, without countermeasures, the calculation of the number of students at the university will provide the incorrect result of three.

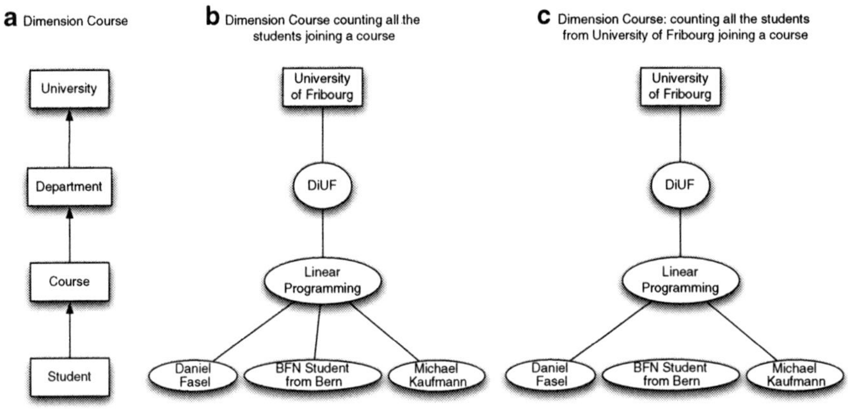

Fig. 2.4 Completeness condition in a dimension "Course"

In addition to disjointness, summarizability defines the condition of completeness. This condition requests that aggregated values on a hierarchy level cover the complete value range of the hierarchy level below it. It is notable to say that completeness is strongly dependent on the type of aggregation respective to its summation operation type. If, for instance, all the students in a course given by the department for informatics are summarized, the final number will contain students from the University of Fribourg but probably also students from another university joining the BFN program. Considering a dimension that aggregates all students in courses and courses to departments, a summation of all the students attending courses from the department or a summation of all students from the university attending the courses can be used as the aggregation function. In the first aggregation all student have to be counted, whereas in the second aggregation the students from external universities can be omitted, even if they join the courses.

Figure 2.4 exemplifies the completeness condition on a dimension "Course", represented in Fig. 2.4a. In Fig. 2.4b the summation of all students joining the course "Linear Programming Language" of the department "DiUF" is represented. Figure 2.4c the summation of students from the "University of Fribourg" joining the course "Linear Programming Language" is shown. Both aggregations fulfill the condition of completeness depending on the type of summation used; Fig. 2.4b summation of all the students and (c) summation of all the students of the University of Fribourg.

The last condition of the summarizability is the type compatibility. Lenz and Shoshani defined different types of measures, which characterize the behavior of the aggregation of these measures over dimension hierarchies. The types of values are categorized in: flow, stock and value per unit type. The type flow can be aggregated by a summation function (SUM). For the type value per unit the SUM aggregation is never allowed. The stock type can be aggregated with SUM as long it is not aggregated over the time dimension. Aggregations by average (AVG), minima

(MIN), maxima (MAX) and count (COUNT) functions are allowed for all types. The following list explains the different types in greater detail:

- **FLOW**: This type of summation allows all possible types of aggregation operations such as summation, product building, maximization, minimization, average building, etc. A possible measure for a FLOW type fact is a quantitative measure like the "Revenue" of a company.
- **STOCK**: This type of summation allows all possible types of aggregation like the FLOW summation type on all dimensions except the dimension time. A possible measure for a STOCK type fact is "amount of articles in a stock".
- **Value Per Unit (VPU)**: This type of summation allows only the aggregation operations: minimization, maximization and average building. A possible measure for a VPU type fact is "customer satisfaction".

2.1.4 Star and Snowflake Schema

There are two well known standards for modeling a multidimensional data structure in a data warehouse. One technique is to use databases with multidimensional data structure storage concepts. These systems are often referred as multidimensional OLAP systems (MOLAP) and some MOLAP systems are commercially used [Ora10]. Because of the native multidimensional storage engine, MOLAP system are interesting for usage as data warehouse [VS00, Vas98]. In reality however, often traditional relational database systems are used as they are much wider spread and are more accepted in business. In order to create a data warehouse based on a relational database system, the multidimensional data aspect has to be modeled on top of classical entities and relationships. Relational databases used as data warehouse are called relational OLAP (ROLAP) systems. Due to the widespread usage of relational database systems, ROLAP systems can be more easily integrated into existing business environments and relational data can be stored more efficiently than in MOLAP systems [VS99]. For projecting multidimensional data structures in relational database systems different modeling techniques have been proposed. The most important modeling approaches are star, snowflake and constellation schema.

The star schema was initially proposed by S. Peterson in 1994 [Pet94]. In a star schema, the data is organized in one fact table and multiple dimension tables. The business measures are stored in the fact table. All dimensional objects, the category attributes, are stored in the dimension tables. For each dimension, one dimension table exists. The fact table and dimension tables are interconnected using a primary – foreign key relation. Figure 2.5 shows a star schema with one fact table and four dimension tables: the dimension "Time" and three generic dimensions. The fact table has a primary key, for each dimension a foreign key and a business measure attribute. The dimension tables have category attributes and a primary key attribute. Each dimension is referenced in the fact table with a primary – foreign key relation.

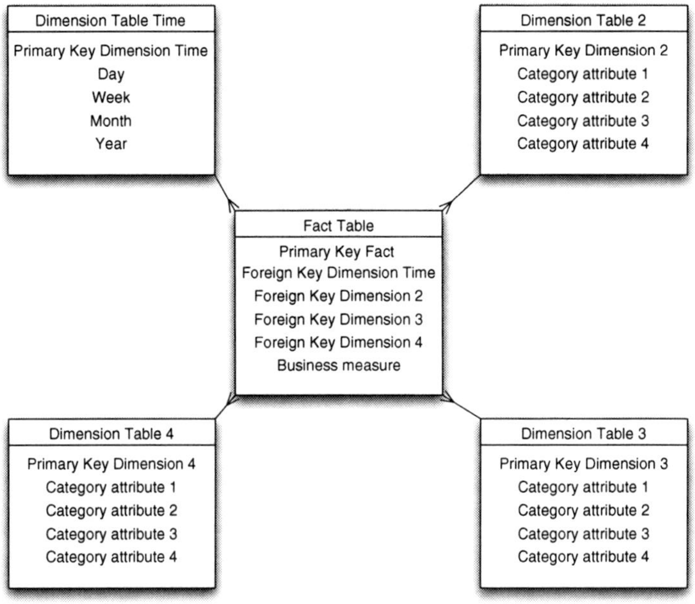

Fig. 2.5 Star schema

The relation dimension to fact is a 1:m relation as a fact instance always has exactly one dimension instance per dimension and a dimension instance can have several fact instances related.

It must be noted that in the star schema, dimension tables are unnormalized. The dimension "Time" demonstrates this unnormalized form of table. A week has 7 days, therefore, every week instance appears seven times in the table. The dimension table is not in the third normal form [Cod71]. The advantage of unnormalized dimension tables lies in the query performance of the data warehouse. Expensive and complex joins over dimension hierarchies can be omitted with unnormalized tables. As dimension tables in general only consume about 5 % of the data storage in a data warehouse and dimension updates are handled centralized [PJ01], normalization is not a crucial element for data consistency in data warehouses.

When the dimension tables of a star schema are normalized, a snowflake schema results. Similar to a star schema, in the snowflake schema the data is organized in a fact table and in multiple dimension tables. But in contrast, for each dimension hierarchy a dedicated table is used. Primary to foreign key relations are used to interconnect the different hierarchies of a dimension. It is not necessary to normalize all the dimension tables. When dimensions are highly unequal in storage size, it might be more convenient to normalize dimensions with big data size and to leave small dimensions unnormalized. Figure 2.6 shows a snowflake schema of a data warehouse. Dimension "Time", "3" and "4" are normalized, whereas dimension "2" is still unnormalized. Between the hierarchy levels of a dimension

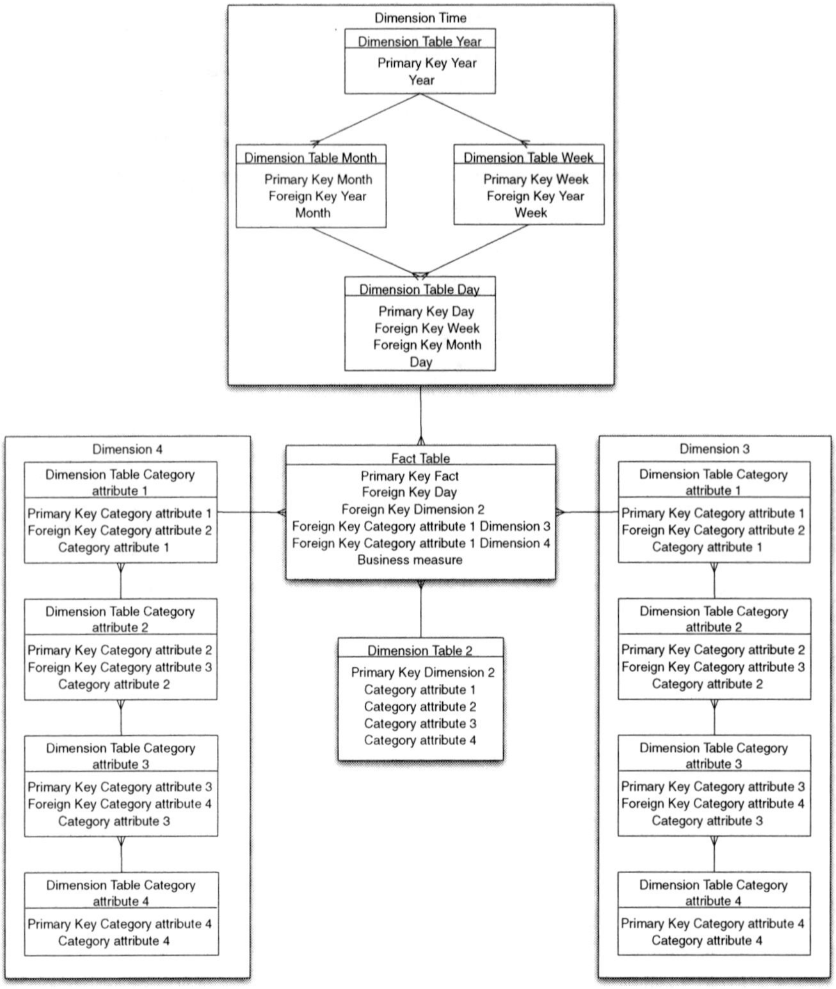

Fig. 2.6 Snowflake schema

1:m relations exist. An instance of a hierarchy level always belongs to exactly one hierarchy instance on the next higher level. On the other hand, an instance from a hierarchy level is composed of one or more instances from the next lower level. This relation is a logical consequence of summarizability (see Sect. 2.1.3). In some cases complex – complex relations might exist in dimension hierarchies. To solve this issue, bridge tables as described by Kimball [Kim98] can be used. The dimension time in this case has two hierarchies: day → week → year and day → month → year.

Despite the fact that normalized dimension tables lead to more complex data warehouse queries, the snowflake schema provides greater flexibility in managing the dimensions because different hierarchies and levels are separated. As long as the referential integrity [Cod87] is maintained, hierarchy levels of complex dimensions can be updated independently [LL03]. In order to optimize query performance, several data warehouse products as DB2 have mechanisms implemented to project normalized dimensions into unnormalized dimensions [Pur02]. The fuzzy data warehouse model described in Chap. 3 uses the snowflake schema. The main intention for using snowflake schema is the possibility to more easily identify elements that have to be classified fuzzily in normalized dimensions.

Kimball et al. [KRT+08] describes one more possible schema for ROLAP systems: fact constellation schema. In this schema, multiple fact tables with different business measures can share the same dimension structure. The advantage of a fact constellation schema is the ability to implement different types of facts and granularities of facts in one data warehouse. This can improve the application area for the data warehouse but complicates the data warehouse model.

2.1.5 Classical Operations

E. F. Codd defined in 1993 [CCS93] the specialized OLAP operations drill-down, roll-up, slice and dice. According Codd et al. [CCS93] and the OLAP Council [OLA10] the OLAP operations, further called classical data warehouse operations, can be described shortly as follows:

- The Roll-up operation consolidates the values of a dimension hierarchy to a value on the next higher level.
- The Drill-down operation is used to navigate from top to bottom in a dimension. It is the opposite operation of the Roll-up operation.
- The Slice operation extracts a subset of values based on one or more dimensions using a dimension attribute to define the subset.
- The Dice operation extracts a subset of values based on more than one dimension using more than one dimension attribute to define the subset. It is an extended version of the Slice operation.

In order to support these operations in ROLAP systems, several extensions for the SQL syntax have been proposed. In the next paragraphs a collection of such SQL extensions are discussed. This discussion will give a first look at how the classical data warehouse operations can be implemented in relational database systems. In a further step, two approaches of defining a multidimensional data cube are presented. These discussions are more general and are not restricted to relational database systems and SQL. Based on the multidimensional data cube approaches, the classical operations are finally defined and discussed in greater detail. These definitions of the classical data warehouse operations are then used as the basis for the extension of the operations in the fuzzy data warehouse.

Gray et al. propose extensions for the GROUP BY clause in SQL Statements [GCB+97]. They propose special cube and roll-up operators in the GROUP

movie	city	year	revenue
	Cit1	Year1	150
Mov1	Cit2	Year1	236
	Cit1	Year2	552
	Cit1	Year1	952
Mov2	Cit2	Year1	157
	Cit1	Year2	555

Fig. 2.7 Possible result set of query in Listing 2.1

BY clause. These new operators enable ordering and aggregating data on multiple dimensions. Agrawal et al. are proposing a similar CUBE BY operator which replaces the GROUP BY clause [AAG$^+$96]. Listing 2.1 shows a SQL statement[1] with the roll-up operator from Gray et al. in which a fact is rolled-up over the dimensions time and region and grouped by the product movie (see Example 1).

Example 1. The example considers a data warehouse with a fact table containing the fact revenue and 3 dimensions region, movie and time. The hierarchies for the dimensions are as follows:

- **Region:** store → city → region
- **Movie:** movie
- **Time:** day → month → year

The dimension tables are named as dim_DIMENSION and are interconnected to the fact table using a join with primary – foreign key relation (e.g. *fact.fk_region* = *dim_region.pk*). The result set (with artificial figures) is a table as shown in Fig. 2.7.

```
select movie, city, "year", sum(revenue) as revenue
from fact
        join dim_region on fact.fk_region = dim_region.
            pk
        join dim_movie on fact.fk_movie = dim_movie.pk
        join dim_time on fact.fk_time = dim_time.pk
group by movie
        roll-up "year"("time") as "year", city(region)
            as city
```

Listing 2.1 SQL example with roll-up operator

[1]The SQL statements in this thesis are in general compliant to the ISO SQL Standard SQL:2008 [ISO08]. However, the database engine used for realizing the fuzzy data warehouse is PostgreSQL 8.4 and therefore some statements might have PostgreSQL specific characteristics.

Fig. 2.8 Possible result set of query in Listing 2.2

city	year	revenue
Cit1	Year1	150
Cit2	Year1	236
Cit1	Year2	552
Cit1	Year1	952
Cit2	Year1	157
Cit1	Year2	555

In Listing 2.2 an SQL statement with the cube operator from Gray et al. in the GROUP BY clause is shown using the same example data warehouse as above. The cube operator aggregates the measure to the hierarchy levels city in dimension region and year in dimension time. A possible result set is shown in Fig. 2.8.

```
select city, "year", sum(revenue) as Revenue
from fact
        join dim_region on fact.fk_region = dim_region.
            pk
        join dim_movie on fact.fk_movie = dim_movie.pk
        join dim_time on fact.fk_time = dim_time.pk
group by cube city(region) as city, "year"("time") as "
    year"
```

Listing 2.2 SQL example with cube operator

Chadhuri et al. analyzed the effectiveness of SQL queries on multidimensional data space. They proposed transformation of the SQL query in order to push the GROUP BY clause in earlier branches of a query tree [CS94]. For SQL views with aggregate function Chadhuri et al. proposed similar improvements [CS96]. A pull-up transformation is proposed for optimizing queries with aggregated views. Additionally they present algorithms that apply these transformations in a cost effective way.

Since around 1998 Microsoft developed the SQL extension Multidimensional Expression language (MDX) which is used in the Microsoft SQL Server product for querying multidimensional data sources [Mic10]. MDX gained a lot of acceptance and is today often considered as quasi standard for SQL extensions for ROLAP systems. Many other products such as the Pentaho Mondrian OLAP Server [Pen10] use MDX as their primary query language.

Other attempts have been made to formulate operations on multidimensional data structure beyond SQL. These attempts follow the notation of multidimensional data cubes and define basic operations on cubes. In this context, basic operation are not the classical data warehouse operations prior mentioned. Basic operations in these approaches are kept more general and the classical data warehouse operation can be derived from them. The advantage of these definitions is the independency

region	movie	time	revenue
Cit1	Mov1	Day1	531
Cit1	Mov1	Day2	548
Cit1	Mov2	Day1	411
Cit1	Mov2	Day2	854
Cit2	Mov1	Day1	813

Fig. 2.9 *R* of basic cube

of a specific query language like SQL. They are therefore more generic and more portable to other technologies such as MOLAP systems. In the next paragraphs, two approaches for multidimensional cubes according Vassiliadis [Vas98] and Agrawal et al. [AGS97] are discussed in greater detail. These approaches are then further used for defining cubes in this thesis.

Vassiliadis proposed a notation of basic cube, cube and multidimensional database as follows [Vas98]:

Definition 4 (Basic cube [Vas98]). A basic cube is as a 3-tuple $< D, L, R >$ where $D =< D_1, \ldots, D_n, M >$ is a list of dimensions D and a measure (a fact) M. $L =< DL_1, \ldots, DL_n, ML >$ is a list of dimension levels DL, aggregated measures ML and R is a set of cells $x = \{x_1, \ldots, x_n, m\}$, where $\forall k \in [1, \ldots, n], x_k \in dom(DL_i)$ and $m \in dom(ML)$ represents the instance values of the basic cube.

For the Example 1, a basic cube may take the following form: $<< region, product, time, revenue >, < city, movie, day, aggregated\ revenue >, R >$ where R is the set of cells represented by Fig. 2.9.

Definition 5 (Cube [Vas98]). A cube is a 4-tuple $< D, L, C_b, R >$ where $D = < D_1, \ldots, D_n, M >$ is a list of dimensions D and a measure (a fact) M. $L = < DL_1, \ldots, DL_n, ML >$ is a list of dimension levels DL and aggregated measure ML. C_b is a basic cube with the restriction $\forall\ d \in C.D\ \exists\ d' \in C_b.D : d = d'$ and R is a set of cells $x = \{x_1, \ldots, x_n, m\}$, where $\forall k \in [1, \ldots, n], x_k \in dom(DL_i)$ and $m \in dom(ML)$ represents the instance values of the cube.

A cube can therefore be denoted as $<< region, product, time, revenue >, < city, movie, month, aggregated\ revenue >, << region, product, time, revenue >, < city, movie, day, aggregated\ revenue > R_b >, R >$ where R_b is represented in Fig. 2.9 and R in Fig. 2.10.

The aim of defining a cube and a basic cube is the traceability of operations. Suppose that a data warehouse operation will calculate the average yearly revenue of movies based on the cube with monthly revenues. It is necessary to go back to daily revenue, the lowest granularity, in order to give a meaningful result on yearly level. If the basic cube of the cube monthly revenue is not known, it is not possible to go to a lower level. No prediction can be made how the daily revenues have been aggregated to monthly revenues. Therefore, one can say that every cube representing

region	movie	time	revenue
Cit1	Mov1	Month1	531
Cit1	Mov1	Month2	548
Cit1	Mov2	Month1	411
Cit1	Mov2	Month2	854
Cit2	Mov1	Month1	813

Fig. 2.10 *R* of cube

a data collection in the data warehouse owns a basic cube. One pure basic cube exists in a data warehouse representing the lowest granularity.

Definition 6 (Multidimensional database [Vas98]). Finally, a multidimensional database is a couple $< D, C >$ where D is a set of dimensions and C is the basic cube representing the lowest granularity.

Another approach for modeling multidimensional cubes is presented by Agrawal et al. [AGS97]. Agrawal et al. proposed a modeling framework in which dimensions and measures are treated symmetrically [AGS97]. The framework defines a multidimensional data cube with a set of basic algebraic operations that are minimal; none of the operations can be replaced with a combination of other operations. The proposed operators show similarities to the relational algebra [Cod70]. It aimed to express functionalities of multidimensional databases as close to relational algebra as possible. Consequently, the operators can be easily projected to SQL statements [AGS95].

Definition 7 (Cube [AGS97]). A cube is composed of:

- k dimensions, with a name D_i and a domain of values dom_i.
- Elements that define the a mapping $E(C)$ from $dom_1 \times \cdots \times dim_k$. The elements are either 0, 1 or a n-tuple. $E(C)(d_1, \ldots, d_k)$ refers to an element at the position $d_1, \ldots, d_k$ in the cube C.
- n-tuple of names describing the n-tuple $< X_1, \ldots, X_n >$ element of C.

If the element of $E(C)(d_1, \ldots, d_k)$ is 0, then no combination of dimensions values exists in the database. One indicates the existence of a value. n-tuples indicates additional information for that dimension combination. One and n-tuple elements can not be intermixed. No distinction is made between measures and dimension. Measures are represented as dimension.

Following Agrawal et al., Fig. 2.11 represents a cube using two dimension product and region. The measure revenue is transformed into a dimension. The elements of the cube are representing the existence of such a dimension combination. Therefore, $E(C)(Day1, 411, Mov2) = 1$ defines the existence of a data entry with revenue $= 411$ for time entry Day1 and article Mov2.

In contrast to the cube definitions of Vissiliadis, Agrawal et al. do not consider the previous state of a cube. Therefore, once an operation is applied to the cube, all

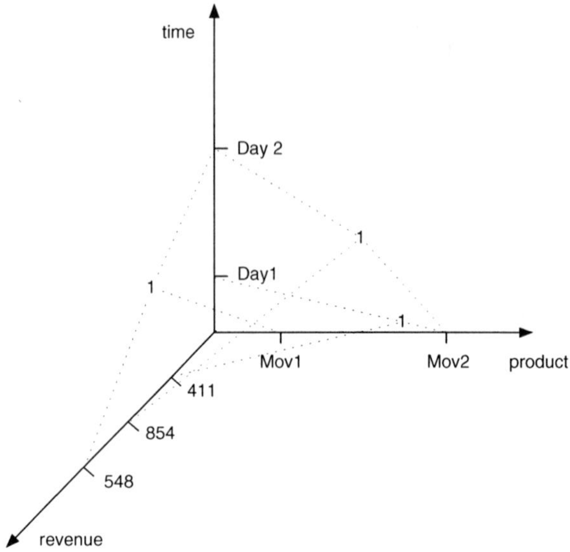

Fig. 2.11 Cube according Agrawal et al. [AGS97]

information of the original cube is modified. For the definition of the Drill-Down operation in Sect. 2.1.5, the previous state of a cube is essential. As a consequence, the characteristics of the cube definitions of Agrawal et al. and Vissiliadis have to combined in the classical data warehouse operation.

Agrawal et al. defined operations for their cube definition. The following two operations from Agrawal et al. are directly used for later definition of the classical data warehouse operations.

Definition 8 (restrict [AGS97]). *restrict*$(C, D_i, P) = C'$ where D_i is the dimension to restrict and P is the predicate defining the restriction in a way that the domain

$$dom_j(C') = \begin{cases} dom_j(C), & if\ 1 \leq j \leq k \wedge j \neq i \\ P(dom_j(C)) \end{cases}$$

and $E(C')(d_1, \ldots, d_k) = E(C)(d_1, \ldots, d_k)$.

The restriction operator removes values of a dimension in a cube that do not satisfy the condition defined in P. Figure 2.12 illustrates on the left side a two dimensional cube with dimensions time and product. When the restriction operation *restrict*$(C, product, \notin \{Mov2\})$ is applied to the cube, all the elements belonging to the dimension value Mov2 are removed. The resulting cube of the restriction operation is illustrated on the right hand side in Fig. 2.12.

Definition 9 (merge [AGS97]). *merge*$(C, \{[D_1, f_1], \ldots, [D_m, f_m]\}, f_{elem}) = C'$ where the dimensions $D_1, \ldots, D_m$ are merged using the corresponding functions $f_1, \ldots f_m$ and the elements are aggregated using the function f_elem in a way that the domain of dimension D_i in C' is

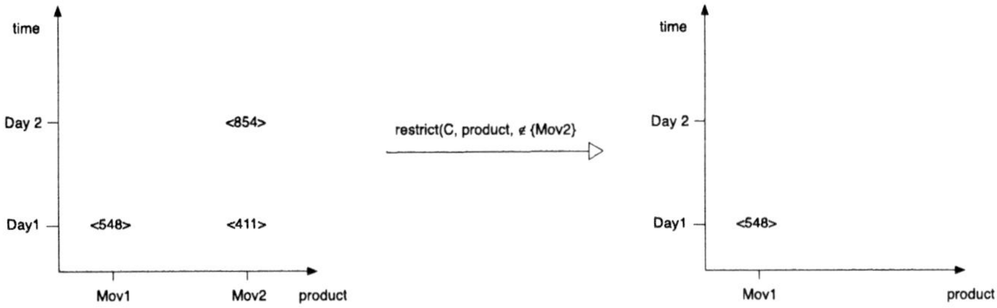

Fig. 2.12 Restriction operation applied to a two dimensional cube

$$dom_i(C') = \begin{cases} f_i(e)|e \in dom_i(C), if\ 1 \leq i \leq m \\ dom_i(C) \end{cases}$$

and $E(C')(d_1, \ldots, d_k) = f_{elem}(\{t|t = E(C)(d'_1, \ldots, d'_k) : f_i(d'_i) = d_i \ \textit{if}\ 1 \leq i \leq m\ else\ d'_i = d_i\})$.

The merge operation is an aggregation function. Values of the merging dimension are combined together to form a new, more condensed value. Multiple merging dimensions can be involved in this process but the combined values are always belonging to the same dimension. For each involved dimension, the merging process is executed separately. The elements are aggregated to the new cube using the f_{elem} function.

Based on the cube and the operation definitions by Vassiliadis [Vas98] and Agrawal et al. [AGS97], the classical OLAP operations roll-up, drill-down, slice and dice can be defined. These definitions are generic enough to be applicable on different technologies and query languages but precise enough to be able to clearly formulate an OLAP operation on a concrete data warehouse example as depicted in Chap. 4.

Example 2. An example data warehouse is used for visualizing the operations. The data warehouse consists of a fact table with a measure revenue and three dimensions as follows:

- **Dimension time** with the hierarchy path day → month → year.
- **Dimension product** with the hierarchy path movie → producer.
- **Dimension region** with the hierarchy path store → city → region.

The snowflake schema of the example data warehouse is presented in Fig. 2.13. The key attributes in the tables are denoted with #PK for primary key and #FK for foreign key.

Every OLAP operation defined below will have the following characteristics:

- The dimension and dimension level are merged using a dot notation into one variable in order to simplify the operation. Therefore, a dimension can be specified as $D = time$ or with including a dimension level as $D = time.month$.

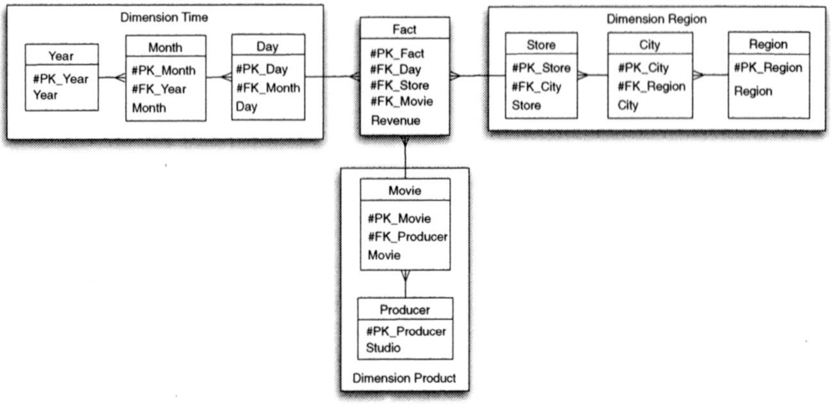

Fig. 2.13 Example data warehouse snowflake schema

- A dimension level always includes the dimension. Therefore, the term dimension level specifies dimension and dimension level.
- Dimension levels in this context are similar to classification attributes as defined in Sect. 2.1.1. The dot notation can be extended in order to integrate other category attributes. A full path to a dimensional attribute month in dimension level month of dimension time can be specified as $D = time.month.month$.
- In the context of the operations, a cube is a data collection of the data warehouse which has the same characteristics as a basic cube defined in Definition 4. As the dimensions D are denoted in the dimension level L, D is redundant and can be omitted (see Definition 4).
- The result set is a new cube. A result cube does not contain any information from the preceding cube or the operation which created the result cube.
- Measures are not directly connected to a dimension or to a dimension level as in Definition 4. As dimension and dimension levels are merged in this cube definition, the measures are extracted from D and denoted as independent element M. The definition of a cube is adapted as follows:

Definition 10 (Cube). A basic cube is a 3-tuple $< D, M, R >$ where

- $D =< D_1, \ldots, D_n, >$ is a list of dimensions, dimensions levels including the dimension or other category attributes including dimension levels and dimensions separated by a dot.
- An aggregated measure (a fact) M
- R is a set of data tuples $x = \{x_1, \ldots, x_n, m\}$, where $\forall k \in [1, \ldots, n] : x_k \in dom(D_k)$ and $m \in dom(M)$ representing the instance values of the cube.

Constraint: Every element in D has to be of the same granularity (either dimension, dimension level or category attribute). If a dimension has a granularity of a category attribute and another dimension of dimension level, the first category attribute of the dimension level is assumed.

Roll-Up

The roll-up operation is used to navigate a dimension upwards. For the dimension region a roll-up operation is executed when the dimension level store is aggregated to the next higher level city.

Agrawal, Gupta and Sarawagi [AGS97] defined a roll-up operation as a special case of their merge operation that it is executed on one dimension. Hence, a roll-up operation can be defined as merging a data cube C to a data cube C′ over one dimension D. In this context a cube has to be understood as a data collection in the data warehouse. An example would be a cube C that is a subset representing the daily revenues; the revenues aggregated to the lowest hierarchy level of dimension time. Cube C can then be merged into a cube C′ that is a data set representing the monthly revenues. This example is a roll-up operation of the measure revenue over the dimension time.

In order to merge, see Definition 9, from one cube to another over one dimension, a function defining how the dimension levels are merged and a function defining how the measures are aggregated have to be specified.

Definition 11 (roll-up). $roll-up(C, D', f_D, f_m) = C'$ where C and C' are cubes (Definition 10), D' is the dimension level to which is merged, f_D is the dimension merge function and f_m is the measure aggregation function.

The domain (dom_D) of D is a set containing the dimension category attribute instances d. $dom_{D'}(C')$ is calculated by applying the function f_D on the dimension category attribute instances d of $dom_D(C)$:

$$dom_{D'}(C') = \{f_D(d) = d' | d \in dom_D(C)\}.$$

The measure $m_{C'}(d)$ of each instance d is calculated by applying the function f_m to each $m_C(d)$ in regards to the aggregation function f_D.

$$m_{C'}(d) = f_m(\{t | t = m_C(d) \wedge f_D(d) = d'\}).$$

$dom_{time.month}(C')$ might be the set $\{January, February\}$ in Example 2. The instances of $dom_{time.month}(C')$ are composed of instances of $dom_{time.day}(C) = \{January1, \ldots, January31, February1, \ldots, February28\}$. The function f_D defines $\{January 1, \ldots, January31\} \rightarrow January$ and $\{February1, \ldots, February28\} \rightarrow February$. A operation $roll-up(< time.day, revenue, R >, time.month, f_d, SUM)$ roll-up the daily revenue to monthly revenue.

The revenue of January 1 might be $m_C(January1) = 8{,}000.-$ and January 31 $m_C(January31) = 5{,}000.-$ in the example. The function f_m is defined as summation and therefore the revenue of January is calculated as $m_{C'}(January) = 8{,}000.- + 5{,}000.- = 13{,}000.$

In a ROLAP based data warehouses, function f_D is implicitly given by the data structure of the dimension tables and the primary – foreign key relations. The function f_m has to be specified explicitly in the query statement or in a meta schema of the data warehouse that allows defining such properties in order to automatically generate appropriate query statements.

Drill-Down

With a Drill-down operation, values in a dimension level will be decomposed in values of the lower dimension level. This operation is used to reveal more detailed levels of data in a dimension. Take, for example, a dimension product in which all the movies are summarized in producer of the movie, this dimension is composed of two hierarchy levels: movies and producers. The drill-down operation in this case will provide a breakdown from a producer to its corresponding movies.

Drill-down is the opposite operation of roll-up. To be able to perform a drill-down, how the category attribute instances are compound from the lower hierarchy level instances must be known in advance. Considering yearly revenue in Example 2, the revenue of the year 2010 can not drilled-down to monthly revenue, if the revenue of every month in 2010 is not known in advance. Otherwise, one can decompose the revenue of 2010 in infinite ways. A roll-up operation defining the aggregation functions and the value instances of the higher hierarchy level has to be executed before a drill-down operation. Hence, the drill-down operation can be considered as a binary operation and formally be defined as:

Definition 12 (drill-down). $drill-down(C, D', f_D^{-1}, f_m^{-1}) = C'$ where C and C' are cubes according Definition 10. D' is the dimension level to decompose, $C = roll-up(C', D, f_D, f_m)$ and f_D^{-1} respectively f_m^{-1} are the inverse function of f_D respectively f_m.

Slice

The slice operation cuts out a slice from a data cube in the multidimensional space of a data warehouse. For example, the cube $C =<< time.year, product.producer, region.city >, revenue, R >$ can be sliced using the value 2010 for the dimensional attribute year. This will extract the revenue by all producer achieved in 2010 in each city. A slice operation can formally be defined as,

Definition 13 (slice). $slice(C, d_m) = C'$ where C is a cube according to Definition 10 and $d_m \in dom(D_m)$ is the element instance that slices the cube.

For extracting the revenue of all producer in all cities in 2010, the slice operation would be defined as follows:

$$slice(C =<< time.year, product.producer, region.city >, revenue, R >, time.year.year = 2010).$$

The slice operation is similar to the inverse restriction operation in Definition 8.

Dice

The Dice operation cuts out a dice from a data cube in the multidimensional space of a data warehouse. Dice is therefore an extended version of the slice operation as it is a slice with multiple slicers. Slicers in a dice are combined using the logical operations AND, OR or NOT. The dice operation can formally be defined as follows:

Definition 14 (dice). $dice(C, \{d_m, \dots, d_k\}, \{f_m, \dots, f_{k-1}\}) = C'$ where C is the cube, $\forall n \in \{m, \dots, k\} : d_n \in dom(D_n)$ are the element instances that slice the cube and $\forall x \in m, \dots, k-1 : f_x \in \{AND, OR, NOT\}$ are the logical operator that combine the slicers in a way that $d_m\ f_m\ d_{m+1}, \dots, d_{k-1}\ f_{k-1}\ d_k$.

As an example of a dice operation, the cube C =<< *time.year, time.month, product.movie, region.city* >, *revenue,R* > can be diced in order to show the revenue of all movies in city Fribourg in February 2010: *dice(C* =<< *time.year,time.month, product.movie,region.city* >, *revenue, R*>, {*time.year.year* = 2010,*time.month. month* = "*February*", *region.city.city* = "*Fribourg*"}, {*AND, AND*}).

Analogous to the slice operation, the dice operation corresponds to the inverse restriction operation in Definition 8.

2.2 Concepts of Fuzzy Logic

This section presents the fundamental concept of fuzzy logic. The fuzzy logic is an application of the fuzzy set theory presented in Sect. 2.2.1 that was first introduced by Zadeh in 1965 [Zad65]. Next, Sect. 2.2.1 presents a set of fundamental operations on fuzzy sets. Based on the fuzzy set theory, the ambiguity of natural language can be mathematically represented by linguistic variables. The concept of linguistic variables is presented in Sect. 2.2.2. Section 2.2.3 finally provides an overview how fuzzy logic is applied in various domains.

2.2.1 Fuzzy Set Theory

Zadeh first introduced fuzzy sets in 1965 [Zad65] as a "class of object with a continuum of grades of memberships". It can be used to mathematically represent ambiguity of classification. For instance, movies might be classified according their genre. A movie can be a romance movie, a comedy, an action movie or a thriller. In this classification it is not always possible to assign a movie to a single class. Often, movies belong to multiple classes at the same time but with different belonging. The movie "Lethal Weapon (1987)" is an action and thriller movie but also contains comedy scenes. Consequently, the movie can not be classified biunique into one set of the classification. In order to truly represent this ambiguity in classification the fuzzy set theory can be applied.

Definition 15 (Fuzzy Set according [Zad65, Zad75b, ZZ80]). A fuzzy set A in $M = \{m\}$, also called universe of discourse [Zad75b], is characterized by a function $\mu_A(m)$ that associates to every element in M a number in the interval $[0, 1]$. The numbers of the interval $[0, 1]$ defines the belonging of element m to the fuzzy set A where 1 implies full belonging and 0 implies no belonging at all. A fuzzy set A in M can be represented as an ordered set of tuples $\{(m, \mu_A(m))\}$.

Example 3. The movie "Lethal Weapon (1987)" (m_1) of the set of all movies (M) can belong to a certain degree to the fuzzy set "action movie" (A). The degree of belonging is specified by the membership function $\mu_A(m_1)$.

Considering the set of movies {"*Lethal Weapon* (1987)″,"*Spaceballs* (1987)″, "*Wall Street* (1987)″}, the fuzzy set "action movie" might contain the following tuples:

"*action movie*″ = {("*Lethal Weapon* (1987)″,1),("*Spaceballs* (1987)″,0.6),("*Wall Street* (1987)″, 0)}.

The movie "Lethal Weapon (1987)" fully ($\mu_{action\ movie}$(*Lethal Weapon* (1987)) = 1) belongs to the fuzzy set "action movie". Whereas, the movies "Spaceballs (1987)" and "Wall Street (1987)" only belong to 0.6, respectively do not belong at all, to "action movie".

Fuzzy Set Operations

The fuzzy set theory is extended with definitions for set theoretic operations. Zadeh [Zad65] first defined basic operations. Over time, other authors have suggested additional and alternative operations as in [DP80, DP85, Ham78, Yag80, Wer88, ZZ80, Zim91]. The following definitions provide an overview of a selection of fundamental fuzzy set operations and characteristics in order to provide a general understanding of fuzzy set theory. Furthermore, different types of set operations that combine fuzzy sets are presented. These set operations are presented in regard to the aggregation of fuzzy concepts later described in Sect. 3.4.2.

Definition 16 (Empty set [Zad65]). A fuzzy set is empty if the membership degrees of every tuple is 0.

Definition 17 (Equal sets [Zad65]). Two fuzzy sets A and B are equal if the membership degrees for every element m in the discourse of universe m is equal for fuzzy set A and B.

$$A = B \text{ if } \forall\ m \in X : f_A(x) = f_B(x).$$

Definition 18 (Complement [Zad65]). The complement A' of a fuzzy set A is a fuzzy set containing the membership degrees $1 - \mu_A(m)$ for every element m in set M. Therefore, the membership function of A' can be defined as:

$$\mu'_A(m) = 1 - \mu_A(m) \; \forall \; m \in M.$$

Definition 19 (α-level set [Zim91]). A set $A_\alpha$that belongs to a fuzzy set A with at least to a degree α is called α-level set:

$$A_\alpha = (m \in M : \mu_A(m) \geq \alpha).$$

An α-cut is consequently an operation that selects an α-level set from a fuzzy set.

When regarding the fuzzy set "action movie" defined in Example 3 an α-level set "*action move*$''_{0.4}$ could be extracted by applying an α-cut of 0.4 to the fuzzy set. The resulting fuzzy set is then as follows:

"*action move*$''_{0.4}$ = {("*Lethal Weapon* (1987)$''$, 1), ("*Spaceballs* (1987)$''$, 0.6)}.

Definition 20 (Convexity of a fuzzy set [Zad65, Zim91]). A fuzzy set A is convex if:

$$\mu_A(\lambda m_1 + (1-\lambda)m_2) \geq min(\mu_A(m_1), \mu_A(m_2)), \text{ where } m_1, m_2 \in M, \lambda \in [0, 1].$$

Alternatively, it can be said that a fuzzy set is convex if all α-level sets are convex.

Definition 21 (Cardinality of a fuzzy set [Zad96, Zim91]). The cardinality of a fuzzy set A is defined as:

$$\Sigma Count(A) = \sum_{m \in M} \mu_A(m)$$

The relative cardinality can consequently be defined as:

$$||\Sigma Count(A)|| = \frac{|\Sigma Count(A)|}{|M|}$$

Different set operators have been proposed for combining fuzzy sets. These operators can be grouped into three main categories: triangular norms (t-norm), triangular conorms (t-conorms) and averaging operators. According to Dubois and Prade, t-norm and t-conorm can be defined as follows [DP85, DP80]:

Definition 22 (t-norm [DP80, Zim91]). t-norm operators are associative, commutative and monotonic two-valued functions that map $[0, 1] \times [0, 1]$ into $[0, 1]$. The following conditions have to be satisfied by the t-norm operator t:

- $t(0, 0) = 0$; $t(\mu_A(m), 1) = t(1, \mu_A(m)) = \mu_A(m)$, $m \in M$
- $t(\mu_A(m), \mu_B(m)) \leq t(\mu_C(m), \mu_D(m))$, if $\mu_A(m) \leq \mu_C(m)$ and $\mu_B(m) \leq \mu_D(m)$
- $t(\mu_A(m), \mu_B(m)) = t(\mu_B(m), \mu_A(m))$
- $t(\mu_A(m), (\mu_B(m), \mu_C(m))) = t((\mu_A(m), \mu_B(m)), \mu_C(m))$

The t-norm operators generally define intersections of fuzzy sets.

Definition 23 (t-conorm [DP85, Zim91]). t-conorm operators are associative, commutative and monotonic two-valued functions that map $[0, 1] \times [0, 1]$ into $[0, 1]$. The following conditions have to be satisfied by the t-norm operator s:

- $s(1, 1) = 1$; $s(\mu_A(m), 0) = s(0, \mu_A(m)) = \mu_A(m)$, $m \in M$
- $s(\mu_A(m), \mu_B(m)) \leq s(\mu_C(m), \mu_D(m))$, if $\mu_A(m) \leq \mu_C(m)$ and $\mu_B(m) \leq \mu_D(m)$
- $s(\mu_A(m), \mu_B(m)) = s(\mu_B(m), \mu_A(m))$
- $s(\mu_A(m), (\mu_B(m), \mu_C(m))) = s((\mu_A(m), \mu_B(m)), \mu_C(m))$

The t-conorm operators generally define unions of fuzzy sets.

Alsina defined in [Als85] the mapping function which can transform any t-conorm function into a t-norm:

$$t(\mu_A(m), \mu_B(m)) = 1 - s(1 - \mu_A(m), 1 - \mu_B(m))$$

The following paragraphs describe the definitions of Zadeh's union and intersection operation. The union function is of type t-conorm and uses a maximum function for combining fuzzy sets. Whereas, the intersection function is a t-norm function and uses the minima operation in order to combine the fuzzy sets.

Definition 24 (Union [Zad65]). The union of two fuzzy sets $A \cup B$ results in a third fuzzy set C with the following membership function:

$$\mu_C(m) = max[\mu_A(m), \mu_B(m)], \text{ where } m \in M.$$

Definition 25 (Intersection [Zad65]). The intersection of two fuzzy sets $A \cap B$ results in a third fuzzy set C with the following membership function:

$$\mu_C(m) = min[\mu_A(m), \mu_B(m)], \text{ where } m \in M.$$

Example 4. In order to exemplify the union and intersection operation on fuzzy sets, another fuzzy set "thriller movie" is added to Example 3, which contains the following tuples:

"thriller movie" = {("*Lethal Weapon* (1987)",0.7),("*Spaceballs* (1987)",0.0), ("*Wall Street* (1987)", 1)}.

The union of the fuzzy sets "action movie" and "thriller movie" is as follows:

"action movie" ∪ *"thriller movie"* = {*max*[("*Lethal Weapon* (1987)",1), ("*Lethal Weapon* (1987)", 0.7)], *max*[("*Spaceballs* (1987)", 0.6), ("*Spaceballs* (1987)", 0.0)], *max* [("*Wall Street* (1987)", 0), ("*Wall Street* (1987)", 1)]} = {("*Lethal Weapon* (1987)", 1), ("*Spaceballs* (1987)", 0.6), ("*Wall Street* (1987)", 1)}

The intersection of the fuzzy sets "action movie" and "thriller movie" is as follows:

"action movie" $\cap$ *"thriller movie"* $=$ {*min*[("*Lethal Weapon* (1987)", 1), ("*Lethal Weapon* (1987)", 0.7)], *min*[("*Spaceballs* (1987)", 0.6), ("*Spaceballs* (1987)", 0.0)], *min*[("*Wall Street* (1987)", 0), ("*Wall Street* (1987)", 1)]} $=$ {("*Lethal Weapon* (1987)", 0.7), ("*Spaceballs* (1987)", 0), ("*Wall Street* (1987)", 0)}

Other t-norm and t-conorm operators for fuzzy sets have been proposed by Dubois and Prade [DP85, DP80], Hamacher [Ham78], Yager [Yag80], Giles [Gil76] and Mizumoto and Tanaka [MT81].

In contrast to t-norm and t-conorm operators, the averaging operators combine fuzzy sets in a way that the resulting membership degree lies between the minima and the maxima of the membership degrees of the combined fuzzy sets. According to empirical studies of Thole et al. [TZZ79], the averaging operators are close to human perception of combining fuzzy sets. For an overview the averaging operator *fuzzy and*, *fuzzy or* [Wer88] and the *compensatory and* [ZZ80] are defined and then discussed in Example 5.

Definition 26 (*Fuzzy AND* and *Fuzzy OR* [Wer88]). The fuzzy AND operator can be defined as:

$\mu_{and}(\mu_A(m), \mu_B(m)) = \gamma \times min[\mu_A(m), \mu_B(m)] + \frac{(1-\gamma)(\mu_A(m)+\mu_B(m))}{2}$, where $m \in M,\ \gamma \in [0,1]$.

The fuzzy OR operator can be defined as:

$\mu_{or}(\mu_A(m), \mu_B(m)) = \gamma \times max[\mu_A(m), \mu_B(m)] + \frac{(1-\gamma)(\mu_A(m)+\mu_B(m))}{2}$, where $m \in M,\ \gamma \in [0,1]$.

The γ parameter indicates how close the *fuzzy and* operation goes to the strict minima, respectively how close the *fuzzy or* operation goes to the strict maximum. If γ is 1 then the *fuzzy and* is the same operation as the intersection and the *fuzzy or* as the union. When γ equals to 0 then the arithmetic mean of the fuzzy sets is calculated.

An average operator that combines fuzzy sets using a combination of the algebraic product and the algebraic sum is the *compensatory and* defined by Zimmermann and Zysno [ZZ80]. According to empirical studies from Zimmerman and Zysno, this operator performs well for aggregation of fuzzy set in the field of human decision making processes [ZZ80].

Definition 27 (Compensatory and [ZZ80]). The compensatory and can be defined as follows:

$\mu_{Ai,comp}(m) = (\prod_{i=1}^{n} \mu_1(m))^{(1-\gamma)}(1 - \prod_{i=1}^{n}(1 - \mu_i(m)))^{\gamma}$ where $m \in M, 0 \leq \gamma \leq 1$.

The γ operator defines where the compensatory and is located between an algebraic sum and the algebraic product. If γ equals 1 the pure algebraic product is used which corresponds to a logical and. Otherwise, the pure algebraic sum representing a logical or is used.

Example 5. In order to exemplify the *fuzzy and*, *fuzzy or* and *compensatory and* operators the two fuzzy sets "action movie" and "thriller movie" are combined by first using a γ operator of 0.5. For the *fuzzy and* the resulting fuzzy set "fuzzy and action, thriller movie" will become:

$$\text{“}fuzzy\ and\ action,\ thriller\ movie'' = \{(\text{“}Lethal\ Weapon\ (1987)'', \mu_{and}(\mu_{action\ movie}(\text{“}Lethal\ Weapon\ (1987)''), \mu_{thriller\ movie}(\text{“}Lethal\ Weapon\ (1987)''))), (\text{“}Spaceballs\ (1987)'', \mu_{and}(\mu_{action\ movie}(\text{“}Spaceballs\ (1987)''), \mu_{thriller\ movie}(\text{“}Spaceballs\ (1987)''))), (\text{“}Wall\ Street\ (1987)'', \mu_{and}(\mu_{action\ movie}(\text{“}Wall\ Street\ (1987)''), \mu_{thriller\ movie}(\text{“}Wall\ Street\ (1987)'')))\}.$$

$$\mu_{and}(\mu_{action\ movie}(\text{“}Lethal\ Weapon\ (1987)''), \mu_{thriller\ movie}(\text{“}Lethal\ Weapon\ (1987)'')) = 0.5 \times min(1, 0.7) + \frac{(1-0.5)(1+0.7)}{2} = 0.775$$
$$\mu_{and}(\mu_{action\ movie}(\text{“}Spaceballs\ (1987)''), \mu_{thriller\ movie}(\text{“}Spaceballs\ (1987)'')) = 0.5 \times min(0.6, 0) + \frac{(1-0.5)(0.6+0)}{2} = 0.15$$
$$\mu_{and}(\mu_{action\ movie}(\text{“}Wall\ Street\ (1987)''), \mu_{thriller\ movie}(\text{“}Wall\ Street\ (1987)'')) = 0.5 \times min(0, 1) + \frac{(1-0.5)(0+1)}{2} = 0.25$$

$$\text{“}fuzzy\ and\ action,\ thriller\ movie'' = \{(\text{“}Lethal\ Weapon\ (1987)'', 0.775), (\text{“}Spaceballs\ (1987)'', 0.15), (\text{“}Wall\ Street\ (1987)'', 0.25)\}$$

The resulting set for the *fuzzy or* "fuzzy or action, thriller movie" will be calculated accordingly and will become:

$$\text{“}fuzzy\ or\ action,\ thriller\ movie'' = \{(\text{“}Lethal\ Weapon\ (1987)'', 0.925), (\text{“}Spaceballs\ (1987)'', 0.45), (\text{“}Wall\ Street\ (1987)'', 0.75)\}$$

A fuzzy set calculated using the *compensatory and* for combining the two fuzzy sets "action movie" and "thriller movie" will be calculated as described next. Again, the γ operator equals to 0.5.

$$\text{“}comp\ and\ action,\ thriller\ movie'' = \{(\text{“}Lethal\ Weapon\ (1987)'', \mu_{comp}(\text{“}Lethal\ Weapon\ (1987)'')), (\text{“}Spaceballs\ (1987)'', \mu_{comp}(\text{“}Spaceballs\ (1987)'')), (\text{“}Wall\ Street\ (1987)'', \mu_{comp}(\text{“}Wall\ Street\ (1987)''))\}.$$

$$\mu_{comp}(\text{“}Lethal\ Weapon\ (1987)'') = (1 \times 0.7)^{(1-0.5)} \times (1-((1-1) \times (1-0.7)))^{0.5} = 0.837$$
$$\mu_{comp}(\text{“}Spaceballs\ (1987)'') = (0.6 \times 0)^{(1-0.5)} \times (1-((1-0.6) \times (1-0)))^{0.5} = 0$$
$$\mu_{comp}(\text{“}Wall\ Street\ (1987)'') = (0 \times 1)^{(1-0.5)} \times (1-((1-0) \times (1-1)))^{0.5} = 0$$

"comp and action, thriller movie" = {(*"Lethal Weapon* (1987)", 0.837), (*"Spaceballs* (1987)", 0), (*"Wall Street* (1987)", 0)}.

When using a γ operator equal to 1 for all three operations, the resulting fuzzy sets are as follows:

"fuzzy and action, thriller movie" = {(*"Lethal Weapon* (1987)", 0.7), (*"Spaceballs* (1987)", 0), (*"Wall Street* (1987)", 0)} which is the same result as the intersection operation in Example 4.

"fuzzy or action, thriller movie" = {(*"Lethal Weapon* (1987)", 1), (*"Spaceballs* (1987)", 0.6), (*"Wall Street* (1987)", 1)} which is the same result as the union operation in Example 4.

"comp and action, thriller movie" = {(*"Lethal Weapon* (1987)", 1), (*"Spaceballs* (1987)", 0.6), (*"Wall Street* (1987)", 1)}.

Conversely, applying a γ operator that equal to 0 will produce the following results:

"fuzzy and action, thriller movie" = {(*"Lethal Weapon* (1987)", 0.85), (*"Spaceballs* (1987)", 0.3), (*"Wall Street* (1987)", 0.5)} which is the arithmetic mean of the membership degrees.

"fuzzy or action, thriller movie" = {(*"Lethal Weapon* (1987)", 0.85), (*"Spaceballs* (1987)", 0.3), (*"Wall Street* (1987)", 0.5)} which is the same result *fuzzy and* operation.

"comp and action, thriller movie" = {(*"Lethal Weapon* (1987)", 0.7), (*"Spaceballs* (1987)", 0.6), (*"Wall Street* (1987)", 0)}.

When using a γ operator equal to 1, the *fuzzy and* becomes the intersection function as Zadeh defined it. Similar, the *fuzzy or* becomes the union operation. Having a γ operator equal to 0, both operations take the arithmetic mean into consideration. The fuzzy set calculated with the *compensatory and* uses the algebraic products for calculating the membership degrees when γ is 0. The algebraic product is a way to express the logical and, which results in a membership degree between 0 and the lowest membership degree. The algebraic product can therefore be used in place of the min operation used in the intersection operation. A similar behavior can be observed when γ is 1. In this case the algebraic sum, as replacement of the max operator in the union operation, is used for calculation which results in a membership degree between the highest membership degree and 1.

Many more fuzzy set operations have been presented in literature that are not described in this section. For further reading on fuzzy set operations, reference a collection of selected papers of fuzzy set theory [ZKY96], Zimmermann's book [Zim91] and Tilli's book [Til91].

2.2.2 Linguistic Concepts

Zadeh developed on top of the fuzzy set theory a means for mathematically representing natural language [Zad78b]. Therefore, he defined a linguistic variable which has words or sentences as its values [Zad75c, Zad75a, Zad75b]. The values of the linguistic variable, further called linguistic terms, are projected on a universe of discourse. Fuzzy sets are used to define the degree of membership with which a value might belong to a linguistic term. Zadeh defines a linguistic variable as follows:

Definition 28 (Linguistic variable [Zad75a]). A linguistic variable is a quintuple $(X, T(X), G, M, F)$ defined as follows:

- X is the name of the linguistic variable
- $T(X)$ is the set linguistic terms of X
- G represents a syntactic rule that generates the set of linguistic terms
- M is the universe of discourse
- F is a semantic rule that defines for each linguistic term its meaning in the sense of a fuzzy subset on U

Example 6. To illustrate a linguistic variable, consider a set of movies. Each movie has a specific runtime. A universe of discourse can now be created as the set of all runtimes of the movies in minutes. Next, a linguistic variable duration can be defined. Duration is further partitioned in a set of linguistic terms $\{very\ short, short, medium, long, very\ long\}$. Finally, for each movie runtime, belonging to the linguistic term can be defined using a semantic rule. The semantic rule for defining the fuzzy subsets by linguistic terms is also called a membership function. The linguistic variable duration can therefore formally be defined as follows:

$$(X = duration, T(X) = \{very\ short,\ short,\ medium,\ long,\ very\ long\}, G, M = [0, 200], F)$$

Here, it is assumed that the runtime of the movies is from 0 to 200 min.

Figure 2.14 illustrates the distribution of the fuzzy subset (F) over the set of all movie runtimes. Each fuzzy subset (different shaded areas) is defined in its shape by the corresponding membership function. Which fuzzy subsets represents which linguistic term is noted in the figure.

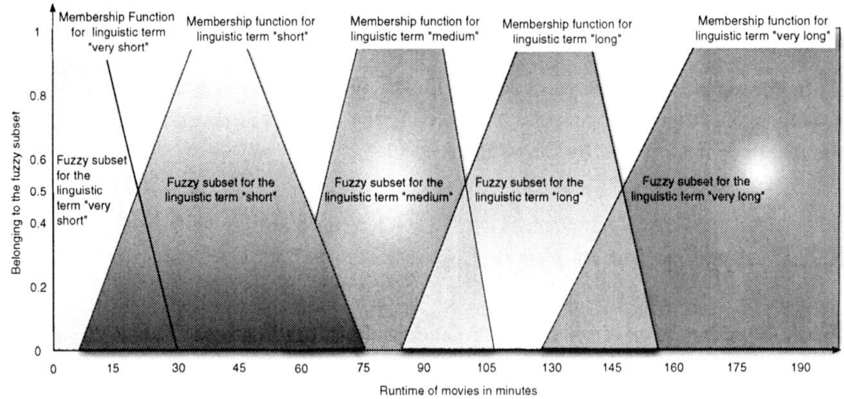

Fig. 2.14 Fuzzy subsets of linguistic terms "very short", "short", "medium", "long" and "very long"

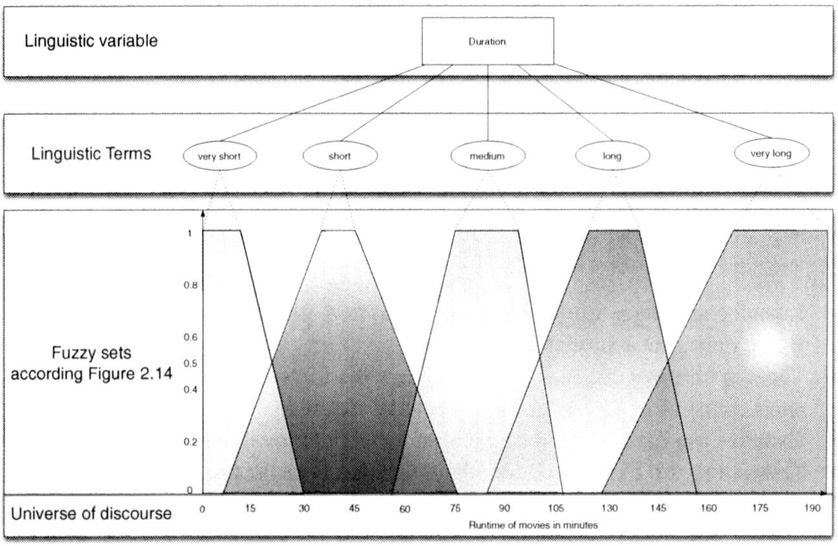

Fig. 2.15 Hierarchical representation of a linguistic variable

The linguistic variable duration represent a fuzzy concept that classifies all runtimes of the movies into its linguistic terms. Figure 2.15 decomposes the linguistic variable duration and its elements to show the hierarchical structure of a linguistic variable. The highest element is the linguistic variable represented by its name "Duration". The linguistic variable is decomposed in its set of linguistic terms. Further, the linguistic terms are projected on the universe of discourse by fuzzy sets.

It is now possible to fuzzily classify movies according their runtime to the linguistic variable duration. For instance, a movie with a runtime of 90 min fully belongs to the class medium duration and belongs with a membership degree of 0.2 to long duration.

In the consecutive chapters of this thesis, the term fuzzy concept is used as a synonym for linguistic variable. The term fuzzy concept is chosen to better denote the usage of fuzzy logic for classification.

2.2.3 *Application of Fuzzy Logic*

Fuzzy logic has been applied to a broad domain of applications. This section gives a short overview of some of these domains.

Soft Computing and Expert Systems

The term soft computing describes different techniques such as probabilistic reasoning, neural networks, fuzzy systems and more. The common purpose of these techniques is that they are all optimized for finding solutions with a certain imprecision and uncertainty [Zad94]. Furthermore, they are adaptive and can change behavior during runtime. Such soft computing applications are often used for controllers that have to constantly react to environmental changes. Tilli [Til91] provides some examples of products that implements fuzzy logic for their functionalities. An excerpt of products described by Tilli [Til91] and where they implement fuzzy logic is provided in the following list:

- Laundry machines use fuzzy logic to control parameters such as quantity of water, detergent and washing time.
- Vacuum cleaners use fuzzy logic to adapt the force of the airflow according the characteristics of the surface that has to be cleaned.
- Cameras use fuzzy logic for their auto focus functionalities.
- Televisions use fuzzy logic for adapting color, contrast and brightness.
- Cars use fuzzy logic amongst other for the anti blocking brake system.

Expert systems according to Kasabov [Kas96] are "knowledge-based systems that contain expert knowledge". They are optimized for solving problems in a specific area and can act similarly as an expert in this domain. Kasabov and Woodford describe in [KW99] how fuzzy logic can be used for expert systems to make them more adaptive and intelligent. In [KW99] it is described how fuzzy neural networks are used for adapting the rule system during runtime in an intelligent expert system.

Fuzzy Possibilities

Zadeh defines possibility theory on the basis of fuzzy sets [Zad78a]. Possibility theory describes how strong the possibility is that an element can be assigned to a fuzzy set. A fuzzy set can then be extended to a possibility distribution. The next example illustrates an application of a possibility distribution.

Example 7. The linguistic term "long" of the linguistic variable duration in Example 6 has a membership function that defines the following fuzzy set (the fuzzy set is not complete but complete enough for an illustration of a possibility distribution):

$$long = \{(80, 0), (90, 0.2), (95, 0.4), (105, 0.7), (125, 1), (130, 1), (145, 0.7)\}$$

Using this linguistic term as a possibility distribution, a proposition such as "Spaceballs is a long movie" can be evaluated. The statement derived from the possibility distribution is that the movie Spaceball might have a runtime of:

- 80 min with a possibility 0
- 90 min with a possibility 0.2
- 95 min with a possibility 0.4
- 105 min with a possibility 0.7
- 125 min with a possibility 1
- 130 min with a possibility 1
- 145 min with a possibility 0.7

In possibility theory the membership degree describes the possibility that an element of the universe of discourse belongs to a fuzzy set. Therefore, it describes the vagueness of the proposition. In contrast, a probabilistic distribution describes the likelihood that a movie has a specific runtime. The difference between possibility and probability theory is discussed by Kosko in greater detail in several publications [Kos90, KI93, Kos94a, Kos94b].

Fuzzy Concepts for Database Applications

Another domain of application is the usage of fuzzy logic in information systems. In [GUP04], Galindo et al. propose new attribute types in relational database systems for representing fuzzy sets. Furthermore, they distinguish between fuzzy data and fuzzy meta knowledge that provides information about the fuzzy attribute types in the database. Galindo et al. propose with this attempt a general framework for implementing fuzzy logic in relational databases.

Other approaches have been proposed that use fuzzy logic for classifying or grouping data in relational database systems [MSW07, MSSV01, MMWS03, MWAS05, MSW08, Wer08]. Schindler [Sch98] and Meier et al. [MSSV01] provide a fuzzy database schema for including linguistic variables in relational databases.

With fCQL they further propose in [Sch98, MSSV01, MMWS03, MWAS05, MSW08] a SQL extension for fuzzily querying the relational database. Werro used this approach for an application on online customer classification in [Wer08].

Fuzzy logic is also applied to multidimensional databases such as data warehouses. These approaches are discussed in greater detail in Sect. 3.1. Data warehouses are often analyzed using data mining techniques in order to generate new knowledge from data. Fuzzy logic can also be applied in the domain of data mining as for instance proposed by Nauck in [Nau00]. Nauck uses neuro fuzzy systems for knowledge discovery and data mining. Kaufmann uses fuzzy induction in [Kau12] for the discovery of relations between attributes in databases.

In Chap. 3, integration of fuzzy logic in data warehouse systems is discussed. Existing approaches are first elaborated on in Sect. 3.1. In the subsequent sections, a new approach is proposed.

Chapter 3
Fuzzy Data Warehouse

In this chapter, a fuzzy data warehouse concept based on meta table structure is presented. To start with, existing approaches are presented and analyzed in Sect. 3.1. The different fuzzy data warehouse approaches are then compared and evaluated in Sect. 3.1.4. In order to address the problems identified in Sect. 3.1.4, Sect. 3.2 presents a new fuzzy data warehouse concept based on a meta table structure. Section 3.2 is divided into four subsections. The first Sect. 3.2.1 discusses basic definition and the meta table definition. Derived from the definitions in Sect. 3.2.1, the fuzzy data warehouse model is defined in Sect. 3.2.2. The definition of the fuzzy data warehouse allows different combinations of meta tables. Therefore, Sect. 3.2.3 provides guidelines that explain how to combine meta tables in order to model different fuzzy concepts. Based on the guidelines in Sect. 3.2.3 and the definition of the fuzzy data warehouse model in Sect. 3.2.2, a fuzzy data warehouse meta model is presented in Sect. 3.2.4. Later, Sect. 3.3 depicts a method which allows transforming a classical data warehouse into a fuzzy data warehouse. This method takes the guidelines and the fuzzy data warehouse meta model into account. Characteristics of fuzzy concepts in a fuzzy data warehouse are presented in Sect. 3.4. Finally, the classical data warehouse operations and fuzzy data warehouse operations are discussed in Sect. 3.5.

Figure 3.1 provides a visual overview of the structure of Chap. 3. The arrows show the direct dependencies of the different sections to each other.

3.1 Existing Research

Numerous efforts to integrate fuzzy logic with data warehouse have been reported. The related work for this study has been divided into three sections. These categories are: data warehouse approach for handling imprecise data (Sect. 3.1.1), approaches for implementing fuzziness into data warehouses (Sect. 3.1.2) and a framework for implementing fuzziness into data warehouses (Sect. 3.1.3). Finally, the related work is further evaluated in Sect. 3.1.4.

D. Fasel, *Fuzzy Data Warehousing for Performance Measurement*, Fuzzy Management Methods, DOI 10.1007/978-3-319-04226-8_3,

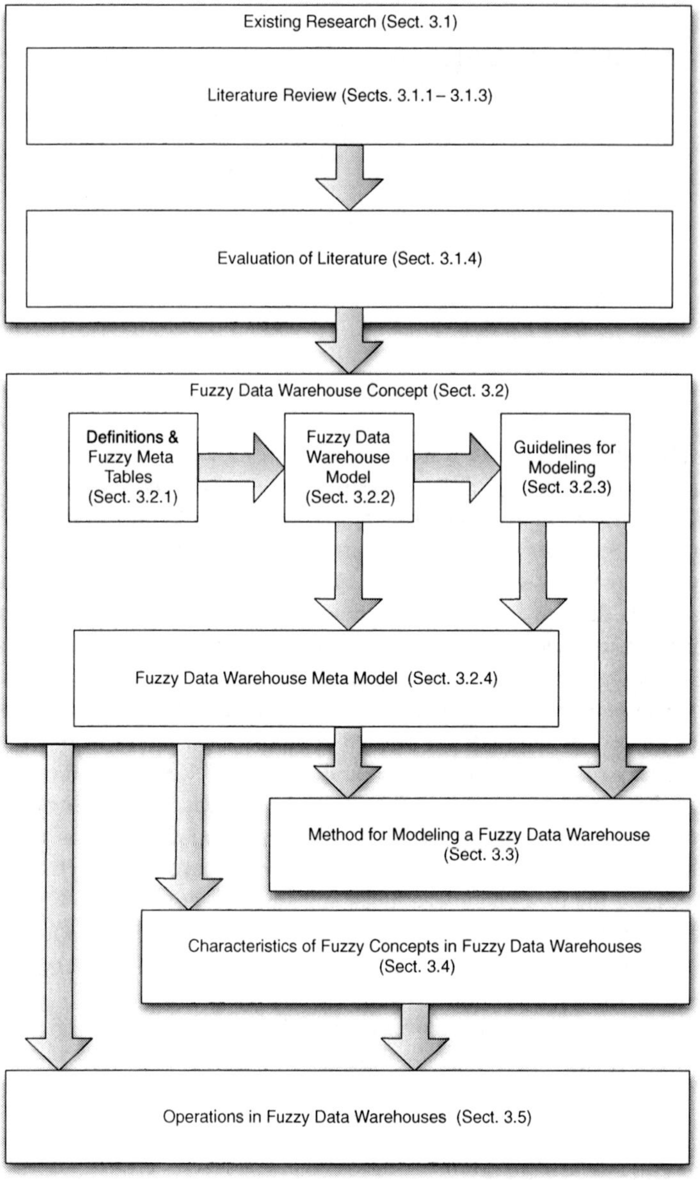

Fig. 3.1 Structure of Chap. 3

3.1.1 *Data Warehouse Approaches for Handling Imprecise Data*

Pedersen, Jensen and Dyreson [PJD99] describe techniques for handling imprecise data in OLAP systems. Pedersen et al. define imprecise data as data with varying granularities. A fact in an OLAP system is often collected from different source systems. These source systems might differ in how they store its measures, for instance a blood pressure value can be stored as a decimal value with precision 2 or with precision 1. OLAP systems in general expect to have all instances of a fact uniform. Therefore data is often merged to the least precise unit. In order to circumvent this drawback, Pedersen et al. allow varying precision in OLAP systems and test before executing a query if imprecise data is involved in the process. In that case, the imprecise data is checked to see if it is still precise enough to answer the query. Otherwise, alternative queries are proposed. The techniques to handle imprecise data are demonstrated by medical data of diabetes patients in [PJD99]. The long time blood sugar level is used to show imprecise data. It might not be available in some data sets or it might be measured with different techniques in different hospitals. The blood sugar is classified as precise data if the value is a decimal number and imprecise data if the value is an integer or unknown. For queries executed on the blood sugar level it is first tested if it can be answered adequately. If the imprecision hinders an adequate answer, an alternative query is proposed. Considering an example of a physician who wants to know all persons having a blood sugar level between 5.4 and 6.2. If the query is executed on data sets that have imprecise blood sugar level measures, the system might propose a query like all persons having a blood sugar level between 5 and 7 instead.

Burdick et al. [BDJ$^+$05] propose an OLAP model to handle both imprecise and uncertain data. According to Burdick et al., imprecision in a OLAP cube occurs when a fact cannot be related with a leaf node instance of a dimension. Leaf node instances are instances of the primary attribute of a dimension. Burdick et al. illustrate an imprecision in OLAP with the example of a car incident in the state Texas. The primary attribute of the dimension location would be city. The fact of the car incident can only be attached to state level and not city level. Therefore, the information about the car incident is considered as imprecise data. In contrast to imprecision, uncertain data are facts that represents likelihood. For instance, a car brake incident might happen with a certain probability on a specific car. Burdick et al. include this likelihood as uncertain fact in the OLAP model. In order to handle this kind of uncertain qualitative information, probability distribution functions are used in the example in [BDJ$^+$05]. The brake incident attribute contains the probability of an incident and for no incident in the following two tuple form: <0.8, 0.2>, with a probability of 0.8 that no brake incident will happen. For aggregating uncertain facts, Burdick et al. propose a linear operator. Furthermore, Burdick et al. explain the aggregation function SUM, AVERAGE, COUNT when combining uncertain and imprecise data and conditions of completeness and faithfulness in order to guarantee the summarizability of the OLAP system.

3.1.2 *Approaches for Implementing Fuzziness into Data Warehouse*

In the area of rules mining on data cubes Alhajj and Kaya [AK03] proposed an implementation of a fuzzy OLAP cube in order to perform fuzzy association rules mining. Based on the classical data warehouse and fuzzy sets, a fuzzy OLAP cube is calculated. The fuzzy OLAP cube contains for each fact the membership degrees of the fuzzy sets as measure. The actual facts of the data warehouse that should be mined are not included in the fuzzy OLAP cube. In a further step, fuzzy association rules are applied to the membership degrees in the fuzzy OLAP cube. Finally, Alhajj and Kaya experimentally prove that this technique produces meaningful results with reasonable efficiency.

Delgado et al. [DMS$^+$04] and Molina et al. [MRASV06] propose integrating fuzzy concepts directly in the dimensional structure of a data warehouse. As an example, a dimension hierarchy of age is depicted. A set of age value 1,...,100 is aggregated over two hierarchy paths age → legal age → all and age → group → all. The dimension attribute legal age has the instances yes and no, whereas, the dimensional attribute group is a fuzzy concept with the linguistic terms young, adult and old as instances. Therefore, all the instances of age are classified within the fuzzy concept groups. When navigating from age to groups, the instances are distributed over the linguistic terms based on their membership degrees. For example, the age 25 might belong to young with a membership degree of 0.7 and to adult with a degree of 0.3. In this case a roll-up to the hierarchy level group would distribute age 25 as 17.5 (0.7*25) to the instance young and 7.5 (0.3*25) to the instance adult.

Schepperle et al. [SMH04] propose a similar approach for integrating fuzzy concepts into dimension structures. Schepperle et al. describe that a fuzzy concept in a data warehouse has the same characteristics as a dimensional attribute and can therefore be aggregated similarly. Conditions for aggregation over fuzzy concepts in dimension structures are further discussed in detail. Based on the summarizability conditions of Lenz and Shoshani [LT01], extended aggregation conditions are proposed. The main outcome of the extended summarizability is the constraint that the membership degrees of all linguistic terms of a fuzzy concept have to be normalized [Sch98]. Additionally, they propose an extension of the Common Warehouse Metamodel [MT02] that covers imprecise data.

Kumar et al. [KKD05] describe an approach for integrating fuzzy concepts into OLAP cubes on facts and dimensions. The fuzzy OLAP cube is defined as a cube that has all facts and at least one dimension fuzzified. In order to fuzzify the facts and the dimensions, the original quantitative values are replaced with a two tuple containing the linguistic term and the corresponding membership degree. The membership degrees are derived using the CLARANS [NH94] clustering algorithm. In Kasinadh and Krishna [KK07] the same approach with multi-attribute summarization for calculating the membership degrees is used. Both approaches essentially integrate fuzzy concepts on dimensions and measures and depict how to operate over them.

Pérez et al. [PSP07] propose a fuzzy spatial data warehouse. The approach is exemplified for a data warehouse representing risk zones around a volcano. In this approach, the fuzzy concepts are applied to the top level attribute of the dimensions. As a result, only the highest aggregated level is fuzzy and the levels beneath are crisp. The fuzzy concept acts as an additional query layer on the crisp data warehouse. By applying fuzzy concepts on the top level attribute of a dimension, the fuzzy concepts are not involved in aggregation operation over the dimension hierarchy. Therefore, the summarizability can be guaranteed as in a classical data warehouse.

Sapir et al. [SSR08] define a method for creating a fuzzy data warehouse. The method depicts how to integrate fuzzy concepts in dimension structures, measures and is based on Kimball's methodology for creating a data warehouse [KR02]. The steps of the method for creating a fuzzy data warehouse are summarized below:

- Identify the business process
- Define the granularity
- Define the dimensions
  - Define the fuzzy dimensions
  - Define the fuzzy hierarchies
- Identify the facts

A fuzzy dimension according Sapir et al. [SSR08] is a dimension that has at least one category attribute with a fuzzy concept. The elements of the fuzzy concept are stored in an additional table and this table is related to the category attribute in a 1:1 relation. The fuzzy concept table comprises the foreign key to the category attribute instance and the linguistic terms as attributes. The membership degrees are the instances of the linguistic terms. A fuzzy hierarchy is a hierarchy in which the category attribute itself is a fuzzy concept and the linguistic terms are the instances of the category attribute. In order to consistently aggregate from the lower hierarchy level to the fuzzy concept, the aggregated values are distributed according the membership degrees. Therefore, an intermediate table between the lower level and the fuzzy concept level is used to hold the membership degrees. A similar concept is known as bridge table in classical data warehouse theory for multivalued hierarchies [KC04, Kim98]. In Example 17, bridge tables are explained in greater detail. Fuzzy concepts on facts are integrated into the fact table as attributes. As a result, for each linguistic term a new attribute has to be specified in the fact table. Additionally, Sapir et al. describe aggregation operations on their fuzzy data warehouse concept.

3.1.3 The Feng and Dillon Framework for Implementing Fuzziness into Data Warehouse

Feng and Dillon [FD03] presented a framework for implementing fuzziness into data warehouse. The framework uses a three layer model to describe the semantics of a fuzzy data warehouse. The first level describes the quantitative view of the data. This level is similar to the classical data warehouse and provides the basic aggregation functions such as SUM, COUNT, AVERAGE, MIN, MAX for the data analysis.

The second layer provides a qualitative summarization of the facts in the data warehouse. Qualitative classification of facts is realized using linguistic terms described as fuzzy term in [FD03]. For each fact instance that has to be classified fuzzily, a set of fuzzy terms and a set of the corresponding membership degrees replaces the quantitative value. Therefore, instead of having a numerical fact in the second layer, the fact instance contains two attributes: the set of fuzzy terms and the set of membership degrees of the fuzzy terms.

The third layer provides quantifier summarization. Feng and Dillon describe fuzzy concepts on dimensions as linguistic quantifiers. A linguistic quantifier is a linguistic probability that determines the degree to which a concept is satisfied [Zad83]. The linguistic quantifier is similar to the fuzzy term in the second layer except that it is applied to the dimension element rather than the quantitative facts of layer 1. The result set of queries on the third layer cube will always provide an overall performance according to a linguistic quantifier of fuzzily grouped (in layer 2) facts.

Further, Feng and Dillon describe the pivot oriented operations slice, dice, roll-up, drill-down and sift applied to their fuzzy data warehouse framework. Each operation can be executed on either a quantitative cube, a qualitative cube or a quantifier cube. Additionally, the measure oriented operations union, intersect and difference are described for each measure type of the cubes. Finally, the way in which cube operations can be transformed into SQL statements is discussed.

3.1.4 Evaluation and Comparison of the Existing Approaches

For comparison and evaluation this section provides a summary of the different approaches and identifies the drawbacks that these approaches have. The approaches are categorized and the drawbacks are depicted. First, the categories differentiate the approaches on whether they apply fuzzy concepts on facts or dimensions. In a second categorization, they are split into approaches that do or do not replace or modify the original values in the data warehouse. Often, the identified problems apply to different categories. Thus, the problems might be discussed multiple times from different perspectives per category. The different approaches are then summarized in a tabular form according to categorization and how the problems can be addressed with the new concept that is presented in Sect. 3.2.

Application of Fuzzy Concepts in Data Warehouse Structure

When regarding where the fuzzy concepts are applied in the data warehouse structure, three different types can be identified. The first type applies fuzzy concepts on the measures of the data warehouse. It either integrates the fuzzy concepts in the fact table or applies them on the aggregated measures of a result set. The following approaches can be identified for applying fuzzy concepts to the facts:

- Pedersen, Jensen and Dyreson [PJD99] handles imprecise facts.
- Burdick et al. [BDJ+05] defines uncertainty on facts by using probabilistic distribution if a certain characteristic (e.g. brake incident) is true or not.
- Alhajj and Kaya [AK03] derives the fuzzy membership degrees from the facts of a crisp OLAP cube and stores them as facts in a fuzzy OLAP cube. The fuzzy OLAP cube is then used for fuzzy association rules mining.

Pedersen, Jensen and Dyreson describe in [PJD99] an approach in which the complexity of handling imprecision is passed to the query engine and further to the user. The user has to decide if the proposed query is adequate for his analysis, which increases the complexity of querying the data warehouse. Imprecise data, as defined by Motro and Smets [MS97], is a special kind of imperfect data by which the accuracy of data is fuzzy. Considering the definition of Motro and Smets, even if the approach in [PJD99] does not directly integrate fuzzy logic into data warehouses, fuzzy classification of imprecise data might facilitate querying such a data warehouse. An alternative solution to the proposition of new queries is to classify blood sugar level with a fuzzy concept containing the linguistic terms low, middle and high. The physician can then query data by using the linguistic terms like: all persons with a high blood sugar level or all persons with at least a 0.8 high blood sugar level. Using a fuzzy data warehouse concept can therefore handle the imprecision of the data during the modeling phase of the data warehouse. Therefore, a major benefit is that the complexity of imprecise data is not passed to the user level and the system does not have to redefine the user query.

Burdick et al. address in [BDJ+05] uncertainty by using likelihoods. Based on Kosko [Kos90], a probabilistic system can be represented by fuzzy theory. The likelihood that brakes are failing can therefore be represented as a normalized linguistic variable with the linguistic terms "no brake incident" and "brake incident". The likelihood value found by the probability distribution can be applied directly as membership degrees to the corresponding linguistic terms in a way that $\mu_{\mathit{no\ brake\ incident}} = 0.8$ and $\mu_{\mathit{brake\ incident}} = 0.2$. Imprecision, as Burdick et al. define, conflicts with the definition of a data warehouse from Inmon, Kimball or Lehner. In all definitions, a key property is the lowest granularity of the information. This property defines that all facts in a fact table can be distinctively related to the primary attribute of a dimension. However, by using a data warehouse model based on a constellation schema as illustrated by Kimball et al. in [KRT+08] it is possible to relate multiple fact tables to the same dimensional structure. With this approach, it might be possible to integrate this kind of imprecise data in a classical data warehouse. For facts with different granularities a distinct fact table can be created and related to the dimensional structure. It is possible to extend a starflake schema to a constellation schema by adding multiple fact tables.

The meta model, which is based on a snowflake model, of the fuzzy data warehouse in Sect. 3.2.4 provides the possibility of adding multiple fact tables in the fuzzy data warehouse model. In conclusion, imprecise and uncertain data as described by Burdick et al. can be interpreted by the fuzzy data warehouse model proposed later in this thesis.

In [AK03], Alhajj and Kaya use the fuzzy facts only for fuzzy association rules mining. This approach limit fuzzy concepts for a specific application scope. The generation of an additional fuzzy OLAP cube can be omitted when directly integrating the fuzzy concepts into the data warehouse. Therefore, the proposed fuzzy data warehouse approach allows fuzzy association rules mining without additional data transformation steps. Due to the ability to simultaneously query crisp and fuzzy data, it might be possible to combine fuzzy association rules mining and other data mining techniques with less effort than with the proposed fuzzy OLAP cube by Alhajj and Kaya. Furthermore with the approach proposed in [AK03], the fuzzy facts are handled in a second OLAP cube, which results in higher costs for storage and synchronization (extraction, transformation and load from crisp cube into fuzzy cube) of the data in the cubes.

Fuzzy concepts are often integrated into the dimension hierarchy. This type of integration then allows aggregation of facts over linguistic terms of the fuzzy concept in dimensions. The following approaches can be identified for applying fuzzy concepts to the dimensions:

- Delgado et al. [DMS$^+$04] and Molina et al. [MRASV06] implement the fuzzy concepts directly into the dimension as additional hierarchy.
- Schepperle et al. [SMH04] propose a similar approach as Delgado et al. [DMS$^+$04] but only provides one fuzzy dimension hierarchy.
- Pérez et al. [PSP07] integrates fuzzy concepts as the top level attribute of the dimensions. Therefore, facts can be aggregated over crisp hierarchy structures and on the last aggregation step they are aggregated into the linguistic terms according their membership degrees.

The approaches from Delgado et al., Molina et al. and Schepperle et al. integrate fuzzy concepts into dimensional structures. This method has several drawbacks. A fuzzy concept reflects a linguistic interpretation or grouping of quantitative values. The interpretation often depends on environmental influences. For instance, specific business terms of a company. Assuming that the linguistic term might change frequently, this approach forces the dimensional structure to adapt more often when using a crisp data warehouse. A second drawback is the need to always have normalized membership degrees for a fuzzy concept. Normalizing membership degrees can lead to information loss. If the membership degree is based on a set of features of a product, normalization removes the information about the features. Example 8 illustrates the difficulty of normalization in greater detail.

Example 8. Swiss army knifes are classified with the fuzzy concept activity. The fuzzy concept activity has the linguistic terms "outdoor", "army","informatics" and "souvenir". The classification is based on the list of features a knife has.

Table 3.1 List of swiss army knife features

Feature	Outdoor	Army	Informatics	Souvenir
Big knife blade	0.3	0.2	0.3	0.1
Small knife blade	0.3	0.2	0.2	0.3
Big knife blade with opening hole	0.4	0.4	0.0	0.0
Screwdriver/can opener	0.3	0.4	0.2	0.1
Phillips screwdriver	0.1	0.2	0.3	0.0
Torx screwdriver no. 2	0.0	0.0	0.4	0.0
Memory stick	0.0	0.0	0.5	0.3
Toothpick	0.0	0.1	0.0	0.2
Tweezers	0.0	0.1	0.0	0.1
Pencil	0.0	0.0	0.2	0.1

Every possible feature gives a certain membership degree in one of the categories. Table 3.1 provides the different features and the corresponding membership degrees for the linguistic term.

The features are combined into different models of swiss army knives. The new swiss army knife has a big knife blade with the opening hole, a screwdriver with can opener and a phillips screwdriver. Therefore it belongs with 1.0 to the class army and with 0.8 to outdoor. An extended version of the new swiss army knife is produced with a small knife blade and a torx screwdriver. The version is now belonging with 1.2 to class army and with 1.1 to outdoor. The membership degree tells how strong the features are met for a specific category. A knife can belong to multiple classes with a membership degree higher 1.0. When normalizing the membership degrees of the fuzzy concept activity, this particular information about the features is lost.

Furthermore, when integrating a fuzzy concept as an additional hierarchy path as proposed in [DMS+04] and [MRASV06], it is not possible to analyze the facts simultaneously crisp and fuzzy. To do so, it is necessary to query the fuzzy data warehouse twice, once using the crisp hierarchy path once the fuzzy hierarchy path.

The fuzzy data warehouse model proposed in Sect. 3.2.2 does not integrate fuzzy concepts in the dimension structure. The fuzzy concepts do not need to be normalized as they are not necessarily involved in the aggregation process of the data warehouse operations (see Sect. 3.5). The fuzzy concepts can be queried in every hierarchy path they are aggregated or propagated on (Sect. 3.4.2). Therefore no specific fuzzy dimension hierarchy is needed. Additionally, the proposed approach also defines fuzzy concepts on facts and not only on dimensions.

The solution from Pérez et al. [PSP07] is developed for a specific application of fuzzy concepts in data warehouses. It does not consider a general approach to integrate fuzziness in data warehouses. Classification with fuzzy concepts is only considered on the highest aggregated level. Beneath this level, it is a classical crisp data warehouse. With the integration of fuzzy concepts in meta tables, as proposed here, the fuzzy spatial data warehouse from Pérez et al. can be modeled. Additionally, this approach is much more generic and can be applied to other domains than just spatial data warehouses.

The third type of approaches describes how to apply fuzzy concepts in facts and dimensions. The manner in which fuzzy concepts on facts can be navigated over fuzzy dimension structures is then shown. The following approaches can be identified for applying fuzzy concepts on dimensions and facts:

- Kumar et al. [KKD05] and Kasinadh and Krishna [KK07] propose the option to fuzzify facts and dimensions. The original crisp values are therefore replaced with the linguistic terms and the corresponding membership values.
- Sapir et al. [SSR08] integrated fuzzy concepts into facts by using an additional fuzzy meta table. Fuzzy concepts on dimension hierarchies can be applied by using a concept similar to bridge tables.
- Feng and Dillon [FD03] propose in their framework two additional OLAP cubes. The first cube integrates the membership degrees of facts as its own facts. The second cube integrates the fuzzy concept on the dimension structures in order to navigate fuzzily over the fuzzy facts of the first cube.

Kumar et al. in [KKD05] and Kasinadh and Krishna in [KK07] replace the quantitative values of the facts and dimensions with the two tuple representing the fuzzy concept. Consequently, information is suppressed and the analysis of this fuzzy data warehouse only provides the qualitative view represented by the fuzzy concept. The membership degrees are automatically defined by clustering algorithms. This approach does not provide the possibility that business analysts can classify manually or that classification is based on external information sources as described in Example 8. The meta table structure of the proposed approach in this thesis does not restrict how the classification is created. It is possible to use automatic classification like clustering algorithms or manual classifications. A classification can be extended with different variants as later explained in Sect. 3.2.3 and therefore this approach is more flexible than the approach proposed in [KKD05].

The approach of Sapir et al. in [SSR08] integrates fuzzy concepts on dimensions in an additional table. This is very similar to the approach in this thesis. However, the structure of the fuzzy concept table differs. In Sect. 3.2.1, it is explained that this approach uses two tables to represent a fuzzy concept. This structure improves the flexibility of the fuzzy concept in two ways. First, the linguistic terms are easier to add, remove or modify. Second, the fuzzy concept can be enhanced with other linguistic term combinations or other classification values as described in Sect. 3.2.3. The integration of fuzzy concepts as a dimension hierarchy is similar to the approaches of Delgado et al. or Schepperle et al. and also suffers from the same drawbacks. The fuzzy concepts defined on facts imply a new attribute for each linguistic term and consequently a new column in the fact table. If a linguistic term is modified or a new fuzzy concept is added, the structure of the fact table has to be adapted accordingly. The fact table is the biggest table in a data warehouse and often holds several million to several billion data rows. It can be very inflexible and dangerous to alter the fact table when a new fuzzy concept is added or an existing fuzzy concept is modified. As a consequence, it will be very difficult to perform changes on fuzzy concepts on facts or adding new fuzzy concepts.

In this thesis, a fuzzy concept on a fact is added within a meta table structure and only related by foreign key relation to the fact table. The fuzzy concept can be modified or deleted without affecting the fact table. This concept provides the same flexibility for fuzzy concepts on facts as on dimensions.

The Feng and Dillon framework in [FD03] differentiates between fuzzy concepts on facts (layer 2) and on dimensions (layer 3). In the approach described in this thesis, this distinction is omitted. As described later in Sect. 3.2.2, fuzzy concepts can be applied to dimensional attributes as they can be applied to facts. The concepts can then be processed similarly and the complexity of transposing from one cube into another is not necessary. The fuzzy data warehouse here proposed also allows the handling of the quantitative crisp values simultaneously with fuzzy concepts. With these functionalities, the fuzzy data warehouse is able to provide the same classification functionalities as described by the framework. Using operations as fuzzifying and defuzzifying described in Sect. 3.5.2 allows integrating and removing fuzzy concepts from a cube in a single step. This is a major advantage compared to the framework as the crisp and the fuzzy values can be compared in a single query. The proposed approach in this thesis already provides the ability to integrate fuzzy concept in the modeling phase of a data warehouse. Snowflake or star schemas can be directly extended with the meta table structures for the fuzzy concepts. This is in contrast to the framework of Feng and Dillon, where the fuzzy concepts are integrated in a second and third layer. Similar to [FD03], classical data warehouse operations like roll-up, drill-down, slice and dice are explained for the fuzzy data warehouse and in Chap. 4, the transformation of the operations in SQL statements are illustrated.

Impact of the Fuzzy Concepts on the Original Values

In a further categorization it possible to distinguish between approaches that replace the original values and approaches that integrate fuzzy concepts in parallel to the original values. In this classification, integrating fuzzy concepts directly in dimension hierarchies is considered as replacement of the original values. This is due to the fact, that the original crisp aggregation path of a dimension is altered. The following approaches do not replace or modify original values:

- Pedersen, Jensen and Dyreson [PJD99] discusse the imprecise nature of the origin facts and do not alter these facts.
- Burdick et al. [BDJ+05] add new uncertain characteristics to the facts, but do not mention the replacement of the original facts.
- Alhajj and Kaya [AK03] create a new fuzzy OLAP cube derived from the crisp OLAP cube. The original OLAP cube remains unchanged.
- Delgado et al. [DMS+04] and Molina et al. [MRASV06] describe the integration of the fuzzy concepts in their own dimension hierarchy. The crisp dimension hierarchy remains untouched.

- Pérez et al. [PSP07] integrates fuzzy concepts directly in the dimension hierarchy but only on the generic top level attribute. Therefore, this integration does not affect the origin aggregation path.
- Feng and Dillon [FD03] provide for each fuzzy concept integration a new cube similar to Alhajj and Kaya [AK03].

The approaches listed below do replace the original values. Either the dimension structure is altered or the fact, respectively dimension, instances are completely replaced by information of the fuzzy concept.

- Schepperle et al. [SMH04] discuss a hierarchy with fuzzy concepts integrated in the dimension. The dimension does therefore not provide an aggregation path without fuzzy concepts involved. It has to be noted that in [SMH04] the fuzzy dimension is never meant to be crisp and is not derived from a crisp dimension hierarchy.
- Kumar et al. [KKD05] and Kasinadh and Krishna [KK07] replace the facts and dimension attributes by the fuzzy membership degrees using algorithms such as CLARANS.
- Sapir et al. [SSR08] integrate fuzzy concepts in the dimensions by using a bridge table and therefore alter the dimension structure. On the other hand, the fuzzy concepts on the facts are placed in a separate attribute and do not replace any fact values.

Summarization of Approaches According Categorization

The following table summarizes the different approaches based on categorization. The stated problems are listed for each categorization and it is shown how the proposed fuzzy data warehouse concept in Sect. 3.2 might be able to overcome the described problems (Table 3.2).

In conclusion, the fuzzy data warehouse model proposed in this thesis is generic enough to cover all the specific solutions discussed in the explored literature. Additionally, most of the discovered problems for integrating fuzzy concepts in data warehouse systems are solved by integrating them into a meta tables structure. Therefore, compared to the discussed approaches, the proposed approach simplifies the integration and the aggregation of the fuzzy concepts and provides a more flexible fuzzy data warehouse.

3.2 Fuzzy Data Warehouse Concept

In order to address the problems described in the Sect. 3.1.4, fuzzy concepts can be integrated as a meta table structure without affecting the core of a data warehouse. The proposed approach is more flexible as it allows integrating and defining fuzzy

Table 3.2 Classification of existing approaches

Categorization	Approaches	Discussed problems	Solution with proposed concept
Fuzzy concepts on facts	• Pedersen, Jensen and Dyreson [PJD99] • Burdick et al. [BDJ+05] • Alhajj and Kaya [AK03] • Kumar et al. [KKD05] and Kasinadh and Krishna [KK07] • Sapir et al. [SSR08] • Feng and Dillon [FD03]	• Automatic calculation of memebrship degrees using algorithms in [KKD05] and [KK07] • Limited application scope in [PJD99], [BDJ+05], and [AK03] • Inflexible for fuzzy concept variants in [SSR08] • High complexity because of multiple OLAP cubes [AK03] and [FD03]	The concept in Sect. 3.2 does not limit how the membership is calculated. Therefore, it allows automatically and manually defined membership functions. The flexibility given by the proposed meta table structure (see Sects. 3.2.2 and 3.2.3) improves this concept for application on different application scopes. Through the use of meta tables structure a single OLAP cube can hold the crisp and fuzzy data and allows analyzing both data at the same time
Fuzzy concepts on dimensions	• Delgado et al. [DMS+04] and Molina et al. [MRASV06] • Schepperle et al. [SMH04]	• When applying fuzzy concepts into dimension hierarchies, the membership degrees have to be normalized as described in [DMS+04], [MRASV06], [SMH04] and [SSR08] • Fuzzy concepts are only applied to top level in [PSP07]	Using the meta table structure as proposed in Sect. 3.2 allows classifying dimension attributes without integrating fuzzy concepts into the hierarchy structure of the dimensions. Therefore, the aggregation of facts over dimension can always The consistency of aggregation is guaranteed and fuzzy concepts do not have to be

(continued)

Table 3.2 (continued)

Categorization	Approaches	Discussed problems	Solution with proposed concept
	• Pérez et al. [PSP07] • Kumar et al. [KKD05] and Kasinadh and Krishna [KK07] • Sapir et al. [SSR08] • Feng and Dillon [FD03]	• Multiple OLAP cubes result in high complexity in [AK03] and [FD03]	normalized. Fuzzy concepts can be integrated on every level of a dimension hierarchy
No replacement or modification of original values	• Pedersen, Jensen and Dyreson [PJD99] • Burdick et al. [BDJ+05] • Alhajj and Kaya [AK03] • Delgado et al. [DMS+04] and Molina et al. [MRASV06] • Pérez et al. [PSP07] • Feng and Dillon [FD03]	• Complexity of imprecision is passed to query engine in [PJD99] • A probabilistic approach is for handling uncertainty in [BDJ+05] • Fuzzy Concepts are normalized in [DMS+04] and [MRASV06] • The complexity for navigation is increased over dimensions in [DMS+04], [MRASV06] and [PSP07] • Multiple OLAP cubes increase complexity in [AK03] and [FD03]	The meta table model proposes an integration of fuzzy concept into an existing data warehouse model as described in Sect. 3.2.4. Consequently, only a single OLAP cube is used for presenting both crisp and fuzzy values. The complexity of synchronizing multiple cube and navigation can be omitted. The complexity of handling imprecise data can be processed with fuzzy concept before querying the data warehouse. Using fuzzy concept instead of a probabilistic approach allows greater flexibility in classification

Replacing or modifying original values	• Schepperle et al. [SMH04] • Kumar et al. [KKD05] and Kasinadh and Krishna [KK07] • Sapir et al. [SSR08]	• The origin data of the data warehouse is altered and therefore it is not possible to provide the same analysis anymore once the fuzzy concepts are integrated	The meta table structure allows querying the data warehouse crisp and fuzzy at once and fuzzy concepts never imply altering a crisp value or structure of the data warehouse. The two additional operations fuzzify and defuzzify are defined in Sect. 3.5 in order to improve the simultaneous query capabilities of the proposed fuzzy concept

concepts without the need for redesigning the core of a data warehouse. By using this fuzzy data warehousing approach, it is possible to extract and analyze the data simultaneously in a classical sharp and in a fuzzy manner. The purpose of this section is to present the definitions of the meta tables, modeling guidelines and the meta model of the fuzzy data warehouse approach.

3.2.1 Basic Definitions and Fuzzy Meta Tables

For integrating fuzzy concepts into a data warehouse, one must first analyze which elements in the data warehouse should be classified fuzzily. The element can be a fact in the fact table or an attribute of a dimension. An element that has to be classified fuzzily is called the target attribute and the value range of the instances of this element is called the domain of attribute.

Definition 29 (Domain of Attribute). A set of possible values or the range of possible values that a dimension attribute or a fact can have is called domain of an attribute or universe of discourse. Domain *Dom* of an attribute *A* is represented by Dom_A.

Definition 30 (Target Attribute). A dimension attribute or a fact that is required to be classified fuzzily is called a target attribute (*TA*). Under fuzzy classification, instances of *TA* are classified over a set (S) that is represented by a linguistic variable. The linguistic variable consists of a set of non-numeric terms called linguistic terms $S = T_1, \ldots, T_k$.

The linguistic terms of a linguistic variable are captured in an attribute called class membership attribute.

Definition 31 (Class Membership Attribute). A class membership attribute (*CMA*) for a target attribute *TA*, represented by CMA_{TA}, is an attribute that has a set of linguistic terms $T_1, \ldots, T_k$ to which the target attribute may belong. In other words, for all possible values of a target attribute (domain of attribute Dom_{TA}) there is a corresponding relation to a *CMA* value. The values of CMA are the values of the set S.

All values of Dom_{TA} belong to a certain fuzzy degree to a *CMA* value. The degree of belonging to a value of *CMA* is called membership degree and this membership degree defines the relation of an instance *TA* to a *CMA* value.

Definition 32 (Membership Degree). The membership degree $MD \in [0, 1]$ is the degree to which the values of a target attribute *TA* are related with linguistic terms $T_1, \ldots, T_k$, respectively with the values of *CMA*.

MD is calculated using a membership function.

Definition 33 (Membership Function). A membership function of a class membership attribute *CMA* is a function $\mu(TA)$ that is used to calculate the membership degree *MD* of a *TA* to a class membership attribute *CMA*: $\mu : TA \rightarrow [0, 1]$

The membership degrees generated by membership functions are captured as membership degree attributes in the fuzzy data warehouse model. A membership degree attribute is defined as:

Definition 34 (Membership Degree Attribute). A membership degree attribute *MDA* of a target attribute *TA*, is an attribute that has a set of membership degrees of the target attribute *TA*. The value of a membership degree is calculated by a membership function and is represented by $\mu_t(TA) = MD$ where *MD* is the membership degree of *TA* for the linguistic term *t* in *CMA*.

An attribute that must be handled fuzzily is extended with two meta tables. The first meta table contains a description of the fuzzy concept and the second meta table contains membership degrees of each instance with regard to the class membership attributes. The two tables are defined as follows:

Definition 35 (Fuzzy Classification Table). A table that consists of linguistic terms and their unique identifiers is called fuzzy classification table *FCT*. It is a two attribute table that consists of an identity attribute and a class membership attribute, where the identity attribute is a unique identifier of the table values. Formally,

$$FCT_{TA} = \{Identifier, CMA_{TA}\}$$

Definition 36 (Fuzzy Membership Table). A table that stores the values representing the degree to which a value is related to a linguistic term is called fuzzy membership table *FMT*. It is a table with four attributes: the identity attribute of the table, the identifier of the target attribute *TA*, the identifier of the class membership attribute CMA_{TA} in the fuzzy classification table FCT_{TA} and membership degree attribute *MDA* for *TA*. Formally,

$$FMT_{TA} = \{Identifier,\ Identifier\ of\ TA,\ Identifier\ of\ CMA_{TA},\ MDA_{TA}\}$$

3.2.2 Fuzzy Data Warehouse Model

The fuzzy data warehouse model is a combination of four types of tables (see Definition 37). These are dimension tables, fact tables, fuzzy membership tables and fuzzy classification tables.

Definition 37 (Fuzzy Data Warehouse Model). A fuzzy data warehouse model is a set of tables and it is represented by *FDW*.

$$FDW = \{Dim,\ Fact,\ FCT_{TA},\ FMT_{TA}\}$$

where,

$$Dim = \{a\ set\ of\ category\ attributes,\ level\ of\ category\ attributes\}$$

$$Fact = \{a\ set\ of\ measures\}$$

$$TA = \{TA_1,\ TA_2,\ \ldots,\ TA_n\}$$

n is the number of attributes that are classified fuzzily.

It is notable that the set of target attributes is a subset of the set of dimension and the set of the facts. Formally, TA is a subset of $Dim \cup Fact$ (i.e. $\forall\ TA_i \in Dim \cup Fact : 1 \leq i \leq n$).

For each attribute $TA_1,\ TA_2\ \ldots,\ TA_n$:

$$FCT_{TA_i} = \{Identifier,\ CMA_{TA_i}\} \text{ where } 1 \leq i \leq n.$$

$$FMT_{TA_i} = \{Identifier,\ Identifier\ of\ FCT,\ Identifier\ of\ TA_i,\ MDA_{TA_i}\} \text{ where } 1 \leq i \leq n.$$

3.2.3 Guidelines for Modeling the Fuzzy Data Warehouse

In this section, a set of guidelines for designing a fuzzy data warehouse model and the usage of these guidelines for developing a meta model for the fuzzy data warehouse is presented.

Distinct Fuzzy Classes/Linguistic Terms

A set of linguistic terms (also called fuzzy classes) is used for the classification of instances of a target attribute. In the simplest case, the linguistic terms are distinct, i.e. there is a single set of linguistic terms with no repetition between them. In this case, one instance of a target attribute belongs to only one fuzzy class at a time and the degree of relation is measured by a membership function. Formally,

$$TA - instance\ (1)\ :\ Fuzzy\ Classes\ (1)$$

Guideline 3.1. *Add a fuzzy classification table FCT and a fuzzy membership table FMT for each target attribute TA, as shown below.*

Different Membership Degrees for the Same Linguistic Terms

An instance of a target attribute may belong to a linguistic term but may have different degrees with which it belongs to a linguistic term. This is due to the fact that multiple business users have different interpretations of a single instance of a target attribute i.e. multiple membership functions are used for a target attribute. Formally,

$$TA - instance\ (1)\ :\ Fuzzy\ Classes\ (1)$$

$$but\ with\ different\ membership\ degrees$$

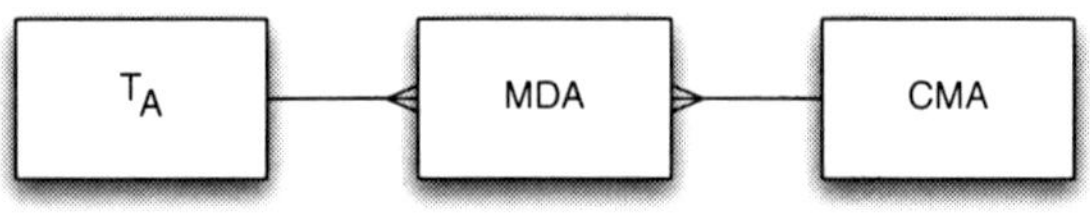

Guideline 3.2. *If an instance of a TA belongs to a fuzzy class but with multiple membership degrees, add a FCT and M number of FMTs, as shown below, where M is the number of distinct membership degrees.*

Different Linguistic Terms for a Target Attribute

An instance of a target attribute may belong to multiple linguistic terms as business users can have more than one set of classes to which an instance of target attribute may belong i.e. multiple fuzzy classes and multiple membership functions. Formally,

$$TA - instance\ (1)\ :\ Fuzzy\ Classes\ (M)$$

Guideline 3.3. *If an instance of a TA belongs to multiple fuzzy classes but with the same membership degree, add M number of FCTs and a FMT, where M is the number of distinct linguistic terms.*

Guideline 3.4. *If an instance of a TA belongs to more than one fuzzy class with different membership degrees, add M number of FCTs and FMTs, as shown below, where M is the number of distinct linguistic terms, and one FCT is related to one FMTs at the most.*

3.2.4 The Fuzzy Data Warehouse Meta Model

According to Harel and Rumpe [HR04], a meta model defines the elements of a conceptualization, as well as their relationships. Figure 3.2 shows the meta model of the proposed fuzzy data warehouse in which the right side shows the meta model of the classical data warehouse. The left side depicts how fuzzy concepts are integrated with a classical data warehouse as a meta tables structure.

The classical data warehouse model class in the meta model refers to a data warehouse schema that is composed of one or more fact tables and two or more dimension tables. A fact table is located at the centre of a data warehouse model and it mainly captures business process measures [KC04]. The relationship with dimension tables is realized with the help of fact attributes. A fact attribute could be a measure (also called fact) or a key attribute (primary key or foreign key). A measure (a subclass of fact attribute) captures critical values about a business process e.g. revenue, whereas a set of key attributes are used to capture the relationship with dimension tables.

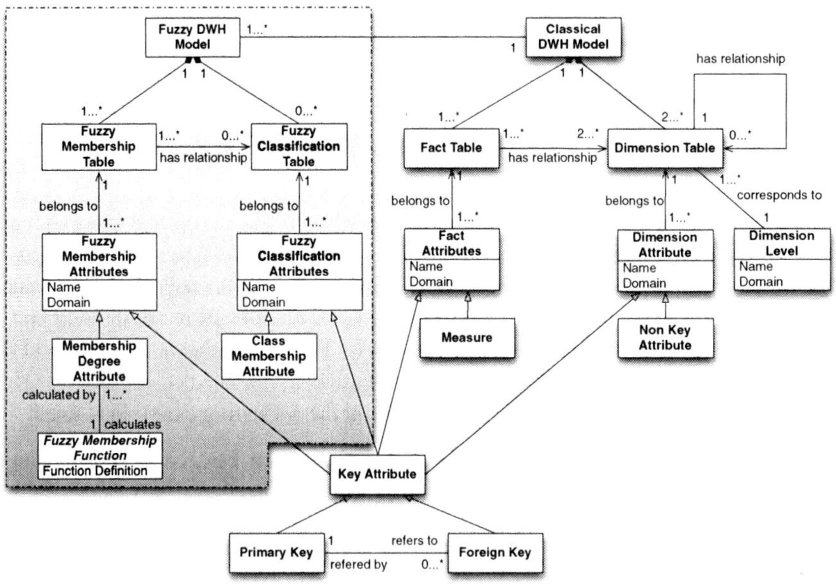

Fig. 3.2 Fuzzy data warehouse meta model

In a classical data warehouse, two or more dimension tables surround a fact table. A dimension table can also be related with one or more other dimension tables to form hierarchies. In this case, each dimension table is at a different hierarchy level (in order to comply with the snowflake schema). The hierarchy level is referred to by the dimension level class in the meta model. A dimension table contains dimension attributes that represent the category attributes or the key attributes of a dimension table. The key attributes capture the relationship between dimension tables, respectively between fact and dimension tables. In addition, other non-key attributes characterize the category attributes of a dimension table.

The fuzzy data warehouse model class in the meta model refers to the fuzzy concept integrated within a data warehouse. For each identified target attribute a fuzzy data warehouse model can be added. Therefore, a classical data warehouse model can have more than one fuzzy data warehouse model. Fuzzy concepts can exist without a linguistic variable. These fuzzy concepts are represented by fuzzy data warehouse models without fuzzy classification table. Every fuzzy classification table has a relation to one or more fuzzy membership table. Therefore, the fuzzy data warehouse model is composed of one or more fuzzy membership tables and zero or more fuzzy classification tables. A fuzzy membership table is built of fuzzy membership attributes. A fuzzy membership attribute might be a key attribute in order to denote primary key or foreign keys. A second type of fuzzy membership attribute is the membership degree attribute. The instances of the membership degree attribute are calculated by the fuzzy membership functions of the fuzzy data warehouse model. The fuzzy classification table contains fuzzy class membership attributes that can be, similarly to fuzzy membership attributes, key attributes.

Additionally, the fuzzy class membership attributes can be a class membership attribute that describes the linguistic term of the fuzzy concept.

3.3 A Method for Modeling a Fuzzy Data Warehouse

In order to create a fuzzy data warehouse, a method is presented that can guide the transformation of a crisp data warehouse into a fuzzy data warehouse. The input to the method is a classical data warehouse and the output is a fuzzy data warehouse. It is a two-step method: in the first step elements of classification are defined and in the second step the fuzzy data warehouse is built. Figure 3.3 shows the tasks and the order in which they are performed.
For illustrating the different steps of the method the following example is used:

Example 9. A data warehouse contains the dimension customer. It is a single hierarchy, single level dimension containing a table customer. Each customer has the attributes name, address and birthday. From the attribute birthday, the age of the customer can be calculated using the function $today - birthday$. Figure 3.4 shows the dimension customer.

3.3.1 Defining Classification Elements

The purpose of this step is to define classification elements that are used in the second step to build the fuzzy data warehouse model. It involves three tasks, identify target attribute, identify linguistic terms and define membership functions. The details are as follows:

First Task: The first task is to identify what should be classified i.e. to identify the target attribute which contains the values that are aimed to be classified fuzzily. This is done by taking into account the end user input. In the simplest case, one target attribute is identified. For Example 9, consider customer age as a target attribute.

Second Task: The second task is to determine how the values of the identified target attribute should be classified i.e. to identify the set of linguistics terms that are used for classifying the instances of a target attribute. Repeat this task for all target attributes. It is represented by iterative loop 1 in Fig. 3.3. There are two possibilities here:

Case 1 – Distinct Linguistic Terms: It is the simplest case in which the linguistic terms are distinct i.e. there is a single set of linguistic terms. Formally,

$$\text{TA instance}\,(1) : \text{Fuzzy Classes}\,(1)$$

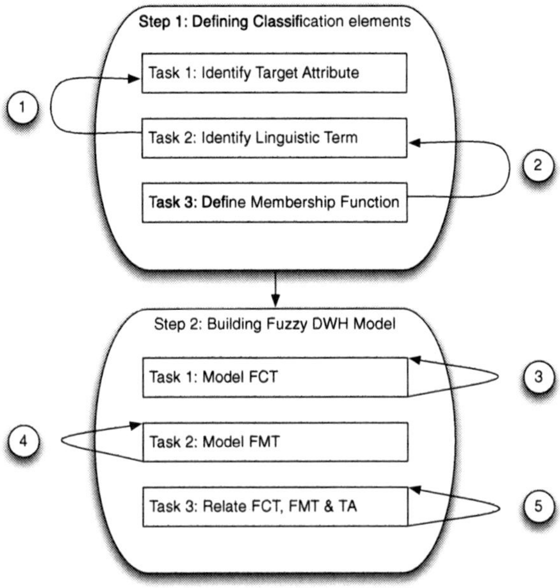

Fig. 3.3 A graphical overview of the method for modeling a fuzzy data warehouse

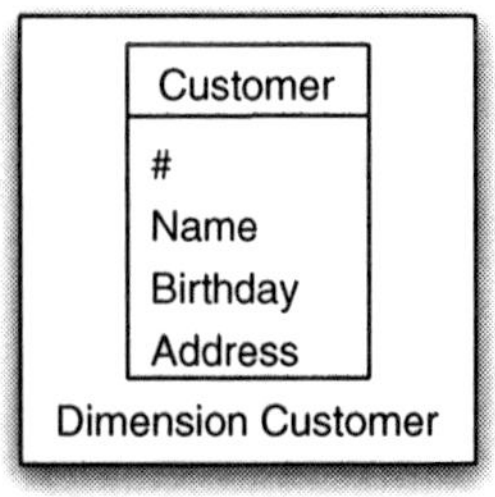

Fig. 3.4 Dimension customer

For the customer age example consider the following set of linguistic terms for {*young age*, *middle age*, *old age*}.

Case 2 – Different Linguistic Terms for a Target Attribute: It is a case in which there are more than one set of linguistic terms for classifying instances of the target attributes. In this case, instances of the target attribute belong to more than one linguistic term as identified by business users. Formally,

TA instance (1) : Fuzzy Classes (M)

For the customer age example consider that the following two set of linguistic terms are identified. These sets are {*young age*, *middle age*, *old age*} and {*teenager*, *adult*, *senior*}. The linguistic terms might already exist in a classical data warehouse model in form of instances of a dimension category attribute. In that case, these terms can be used for classifying instances the of target attribute.

Third Task: The third task is to define a membership function (represented by μ) for each linguistic term. It is done in such a way that the values can be determined over a scale of 0 to 1. Repeat the task for each identified linguistic term. It is represented by iterative loop 2 in Fig. 3.3. It could be the case that for different users a target attribute belongs to the same set of linguistic terms with different membership degrees. The case is as follows:

Case 3 – Different Membership Degrees for the same Linguistic Terms: It is a case in which an instance of a target attribute belongs to a linguistic term with different membership degrees. It can be due to the reason that multiple business users have different interpretations of a single instance of a target attribute i.e. multiple membership functions are used for a target attribute. Formally,

$$\text{TA} - \text{instance } (1) : \text{Fuzzy Classes } (1)$$

but with different membership degrees.

Below, we discuss the examples of the third task for each case:

Example of Task 3 for Case 1 – Distinct linguistic terms. For the customer age example, a membership function is defined for each linguistic term. The membership functions μ_{young}, μ_{middle}, μ_{old} become:

$$\mu_{young}(customer\_age) = \begin{cases} if\ customer\_age \leq 15\ , & MD\_AgeGroup = 1 \\ if\ customer\_age \geq 25\ , & MD\_AgeGroup = 0 \\ else & , MD\_AgeGroup = \frac{25-customer\_age}{25-15} \end{cases}$$

$$\mu_{middle}(customer\_age) = \begin{cases} if\ customer\_age \leq 15 & , MD\_AgeGroup = 0 \\ if\ customer\_age \geq 60 & , MD\_AgeGroup = 0 \\ if\ 25 \leq customer\_age \geq 40 & , MD\_AgeGroup = 1 \\ if\ 15 < customer\_age > 25 & , MD\_AgeGroup = \frac{customer\_age-15}{25-15} \\ else & , MD\_AgeGroup = \frac{60-customer\_age}{60-40} \end{cases}$$

$$\mu_{old}(customer\_age) = \begin{cases} if\ customer\_age \leq 40\ , & MD\_AgeGroup = 0 \\ if\ customer\_age \geq 60\ , & MD\_AgeGroup = 1 \\ else & , MD\_AgeGroup = \frac{customer\_age-40}{60-40} \end{cases}$$

Example of Task 3 for Case 2 – Different Linguistic Terms for a Target Attribute. For case 2 of customer age example, a membership function is defined for each linguistic term i.e. μ_{young}, μ_{middle}, μ_{old}, $\mu_{teenager}$, μ_{adult}, μ_{senior}.

The membership functions μ_{young}, μ_{middle}, μ_{old} are the same as above and $\mu_{teenager}$, μ_{adult}, μ_{senior} become:

$$\mu_{teenager}(customer\_age) = \begin{cases} \text{if customer\_age} \leq 15 \text{ , MD\_AgeGroup} = 1 \\ \text{if customer\_age} \geq 25 \text{ , MD\_AgeGroup} = 0 \\ \text{else} \qquad\qquad\qquad\quad \text{, MD\_AgeGroup} = \\ \qquad\qquad\qquad\qquad\qquad \frac{25-customer\_age}{25-15} \end{cases}$$

$$\mu_{adult}(customer\_age) = \begin{cases} if\ customer\_age \leq 15 & , MD\_AgeGroup = 0 \\ if\ customer\_age \geq 60 & , MD\_AgeGroup = 0 \\ if\ 25 \leq customer\_age \geq 40 & , MD\_AgeGroup = 1 \\ if\ 15 < customer\_age > 25 & , MD\_AgeGroup = \\ & \frac{customer\_age-15}{25-15} \\ else & , MD\_AgeGroup = \\ & \frac{60-customer\_age}{60-40} \end{cases}$$

$$\mu_{senior}(customer\_age) = \begin{cases} if\ customer\_age \leq 40 & , MD\_AgeGroup = 0 \\ if\ customer\_age \geq 60 & , MD\_AgeGroup = 1 \\ else & , MD\_AgeGroup = \\ & \frac{customer\_age-40}{60-40} \end{cases}$$

Example of Task 3 for Case 3 – Different Membership Degrees for the same Linguistic Terms. For case 3 of the customer age example, a membership function for each linguistic term becomes:

$$\mu_{young}(customer\_age) = \begin{cases} \text{if customer\_age} \leq 15 \text{ , MD\_AgeGroup} = 1 \\ \text{if customer\_age} \geq 25 \text{ , MD\_AgeGroup} = 0 \\ \text{else} \qquad\qquad\qquad\quad \text{, MD\_AgeGroup} = \\ \qquad\qquad\qquad\qquad\qquad \frac{25-customer\_age}{25-15} \end{cases}$$

$$\mu_{middle}(customer\_age) = \begin{cases} if\ customer\_age \leq 15 & , MD\_AgeGroup = 0 \\ if\ customer\_age \geq 60 & , MD\_AgeGroup = 0 \\ if\ 25 \leq customer\_age \geq 40 & , MD\_AgeGroup = 1 \\ if\ 15 < customer\_age > 25 & , MD\_AgeGroup = \\ & \frac{customer\_age-15}{25-15} \\ else & , MD\_AgeGroup = \\ & \frac{60-customer\_age}{60-40} \end{cases}$$

$$\mu_{old}(customer\_age) = \begin{cases} if\ customer\_age \leq 40 & , MD\_AgeGroup = 0 \\ if\ customer\_age \geq 60 & , MD\_AgeGroup = 1 \\ else & , MD\_AgeGroup = \frac{customer\_age-40}{60-40} \end{cases}$$

In order to relate the customer age with the same linguistic terms using another membership degree, we define another membership function for each linguistic term i.e. $\mu_{young}1$, $\mu_{middle}1$, $\mu_{old}1$. The definitions of the membership function become:

$$\mu_{young}1(customer\_age) = \begin{cases} \text{if customer\_age} \leq 20 \text{ , MD\_AgeGroup} = 1 \\ \text{if customer\_age} \geq 40 \text{ , MD\_AgeGroup} = 0 \\ \text{else} \qquad\qquad\qquad \text{, MD\_AgeGroup} = \\ \qquad\qquad\qquad\qquad \frac{40-customer\_age}{40-20} \end{cases}$$

$$\mu_{middle}1(customer\_age) = \begin{cases} if\ customer\_age \leq 40 & , MD\_AgeGroup = 0 \\ if\ customer\_age \geq 85 & , MD\_AgeGroup = 0 \\ if\ 55 \leq customer\_age \geq 70 & , MD\_AgeGroup = 1 \\ if\ 40 < customer\_age > 55 & , MD\_AgeGroup = \\ & \frac{customer\_age-40}{55-40} \\ else & , MD\_AgeGroup = \\ & \frac{85-customer\_age}{85-70} \end{cases}$$

$$\mu_{old}1(customer\_age) = \begin{cases} if\ customer\_age \leq 70 & , MD\_AgeGroup = 0 \\ if\ customer\_age \geq 85 & , MD\_AgeGroup = 1 \\ else & , MD\_AgeGroup = \\ & \frac{customer\_age-70}{85-70} \end{cases}$$

3.3.2 *Building Fuzzy Data Warehouse Model*

The purpose of this step is to employ the identified classification elements to build a fuzzy data warehouse model. It involves three tasks, create fuzzy classification table, create fuzzy membership table and relate the tables. The details are as follows:

First Task: The first task towards building a fuzzy data warehouse model is to model a fuzzy classification table for each set of linguistic terms. As described in Sect. 3.2.1, it is a two attribute table in which one attribute is the identifier of the table and second attribute is the class membership attribute. The values of the class membership attribute are the values of the linguistic terms. As stated above, the task should be repeated for each set of linguistic terms, it is represented by the loop 3 in Fig. 3.3.

Second Task: The second task in building a fuzzy data warehouse model is to create fuzzy membership tables. As described in Sect. 3.2.1, it is a four attribute table in which the first attribute is the identifier of the table, the second attribute is the identifier of the target attribute, the third attribute is the identifier attribute of fuzzy classification table and the fourth attribute is the membership degree attribute for the target attribute. The values of membership degree attribute are calculated by membership functions, as identified above.

For the cases described above, we present some guidelines for the first and the second task, which are as follows:

Case 1 – Distinct Linguistic Terms: It is the simplest case in which there is a single set of linguistic terms with no repetition between them. For this case, define a membership degree attribute (MDA) and a class membership attribute (CMA) for a target attribute, as given below:

TA (1) : (1) MDA; MDA (1) : (1) CMA

According Guideline 3.1, add a fuzzy classification table (FCT) and a fuzzy membership table (FMT) for target attribute TA, as given below:

Dimension/Fact (1) : (1) FMT and FMT (1) : (1) FCT

For the customer age example, there is one distinct set of linguistic terms i.e. {*young age*, *middle age*, *old age*}. Therefore, by following Guideline 3.1, add a *FCT(customer age)* = *AgeGroup_ID, CMA_AgeGroup*} and a *FMT (customer age)* = {*Membership_ID, AgeGroup_ID, Customer_ID, MDA_ AgeGroup*}.

Case 2 – Different Linguistic Terms for a Target Attribute: It is a case in which there is more than one linguistic term for a target attribute. For this case, define multiple membership degree attributes (MDA) and multiple class membership attributes (CMA) for a target attribute, where one membership degree attribute corresponds to a class membership attribute, as given below.

TA (1) : (M) MDA and MDA (1) : (M) CMA

Guideline 3.3 describes that if an instance of a TA belongs to multiple fuzzy classes but with the same membership degree, add M number of FCTs and a FMT, where M is the number of distinct set of linguistic terms.

Additionally, Guideline 3.4 describes that if an instance of a TA belongs to more than one fuzzy class with different membership degrees, add M number of FCTs and FMTs, as given below, where M is the number of distinct set of linguistic terms. The addition of M number of FCTs and FMTs is represented by loop 3 and 4 respectively in Fig. 3.3.

Dimension/Fact (1) : (M) FMT and FMT (1) : (M) FCT

For the customer age example, there are two distinct set of linguistic terms i.e. {*young age*, *middle age*, *old age*} and {*teenager*, *adult*, *senior*}. Therefore, following Guideline 3.3 add *FCT1(customer age)* = {*AgeGroup1_ID, CMA_ AgeGroup*} and *FCT2(customer age)* = {*AgeGroup2_ID, CMA_AgeGroup*}, one for each set. Also, add a *FMT(customer age)* = {*Membership_ID, Age-Group1_ID, AgeGroup2_ID, Customer_ID, MDA_AgeGroup*}.

Case 3 – Different Membership Degrees for the same Linguistic Terms: It is a case in which target attributes belong to a linguistic term with different degrees. For this case, define multiple membership degree attributes (MDA) and a class membership attribute (CMA) for a target attribute, as given below.

TA (1) : (M) MDA and MDA (M) : (1) CMA

Guideline 3.2 describes that if an instance of a TA belongs to a fuzzy class but with multiple membership degrees, add a FCT and M number of FMTs, where M is the number of distinct membership degrees.

Dimension/Fact (1) : (M) FCT and FMT (M) : (1) FCT

For the customer age example, there is one set of linguistic terms i.e. {*young age*, *middle age*, *old age*}, however different membership functions are defined. Therefore, following Guideline 3.2 add a *FCT*(*customer age*) = {*AgeGroup_ID*, *CMA_AgeGroup*}, a *FMT*1(*customer age*) = {*Membership ID*, *AgeGroup_ID*, *Customer_ID*, *MDA*1*_AgeGroup*} and *FMT*2(*customer age*) = {*MembershipID*, *AgeGroup_ID*, *Customer_ID*, *MDA*2*_AgeGroup*}.

Third Task: The third task is to relate each fuzzy membership table and fuzzy classification table with the "to be classified" table i.e. the table that contains the target attribute. For that, add a foreign key relation *FMT*(*Identifier of TA*) = *TA* (*Identifier*) to relate the TA table with the FMT table. Second, add a foreign key relation *FMT*(*Identifier of FCT*) = *FCT* (*Identifier*) to relate each FCT and FMT. Figure 3.5 presents the relation of the fuzzy concept tables and the dimension table client. The meta tables for the fuzzy concepts are grey shaded. The table FCT_Customer_Age1 represents *FCT*1 that contains the linguistic terms {*young age*, *middle age*, *old age*} in the cma attribute. Accordingly, FCT_Customer_Age2 represents *FCT*2 that contains the linguistic terms {*teenager*, *adult*, *senior*} in the CMA attribute. Table FMT_Customer_Age1 represents *FMT*1 with the corresponding membership degrees in the MDA. Similar, FMT_Customer_Age2 represents *FMT*2 and FMT_Customer_Age3 represents *FMT*3.

3.4 Characteristics of Fuzzy Concepts in Fuzzy Data Warehouse

Data collection in a data warehouse implies various challenges for using fuzzy concepts. Data is analyzed over time and under different aspects defined by the dimensionality of the data warehouse. The value range of data collection is therefore directly dependent on aspects of the analysis. Because of that, the analysis of data defines the level of aggregation, the dimensions on which data is analyzed and the

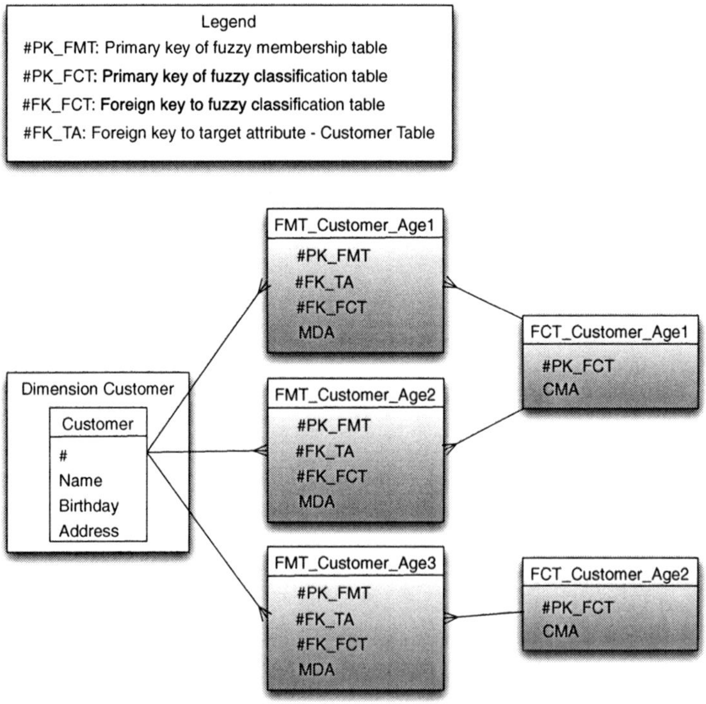

Fig. 3.5 Dimension customer with fuzzy concepts

time frame. A fuzzy concept applied to a data collection as target attribute has to be able to handle these aspects. The next sections will cover different characteristics that a fuzzy concept must possess in order to fully support data collections as target attributes.

3.4.1 *Types of Fuzzy Concepts*

In contrast to a transactional system, data in a data warehouse is populated on regular bases and never deleted and rarely modified. The data size therefore grows steadily and over the years, the value range might grow simultaneously. If the data size grows over the physical capacity of the storage system, it is necessary to partition and to archive parts of it. Consider a movie rental company data warehouse that stores information about movies. It is likely that over time more movies and different genres are added. Data of old movies that are not available anymore in the stores is sorted out and archived separately. Considering this mechanism, one can state that the value ranges of movies and genres are fluctuating over time.

Fuzzy concepts should be flexible in the sense that they should properly handle fluctuating value ranges. Depending on how the fluctuation should be handled,

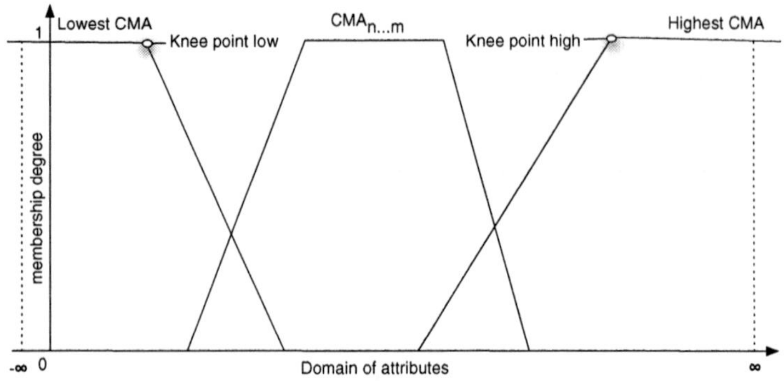

Fig. 3.6 Schematic example of an open end fuzzy concept

different fuzzy concept types can be defined. Fuzzy concepts can be characterized into three types: open ended, limited and adaptive fuzzy concept.

Open End Fuzzy Concepts

The Fig. 3.6 shows a schematic representation of an open end fuzzy concept. The lowest class membership attribute (CMA) and the highest CMA both have a knee point. All target attribute values below the lowest knee point or above the highest knee point will have a membership degree of 1 of the particular CMA.

An open end fuzzy concept can be defined as a fuzzy concept ($\{CMA\_low, \ldots, CMA\_high\}$) for which the domain of the target attribute is infinite. Value instances below the lower knee point (LK) fully belong to the lowest CMA and value instances above the higher knee point (HK) fully belong to the highest CMA. Formally,

Definition 38 (Open end fuzzy Concept).

$$dom_{TA} = [-\infty, +\infty] \ \wedge \ x_{low}, x_{high} \in TA \ \wedge \ x_{low} < LK \ \wedge \ HK < x_{high} \ \wedge$$
$$\mu_{CMA\_low}(x_{low}) = 1 \wedge \ \mu_{CMA\_high}(x_{high}) = 1.$$

With an open end fuzzy concept, the complete domain of target attribute is covered, even if the domain is varying over time. The calculation cost of the open end fuzzy concept is relatively low as target attribute values beyond or above knee points always receives a membership degree of 1. Only for values in between a membership function the membership degree has to be calculated. This kind of fuzzy concept is stable in regard of target attribute values which are not in the standard distribution of the domain of attributes. These outliers are simply classified sharp, respectively with a membership of 1 in either the lowest or the highest CMA. In contrast to the adaptive fuzzy concept these outliers do not affect the classification of the other target attribute value.

On the other hand the fuzzy concept will only give an accurate classification between the knee points. If the value range of the target attributes is growing and therefore most of the values are classified after the knee points, it will be necessary to adapt the concept to the new domain of attributes. This operation might be cost intensive as all the membership degrees of all target attribute values have to be readjusted.

The open end fuzzy concept requires as many membership degree attributes in the fuzzy membership table as the number of the Cartesian product of the class membership attributes and the target attributes. For every new value the membership degree has to be calculated. It should be noted that knee points have to be chosen in a way that distribution of the values in the different class membership attributes reflects the nature of the classification. If the value range increases over time and the knee points are chosen to close together, it is probable that values fully belong to easily to either the lowest or highest class membership attribute.

For illustration of the different characteristics of fuzzy concept consider the following example:

Example 10. A data warehouse contains two dimensions: client and employee. Both dimensions contain the dimensional attribute birthday from which the age can be calculated. Additionally, both dimensions have the dimensional attribute name and the customer dimension has the dimensional attribute address. The dimensions are single hierarchy and single level dimensions and are realized by one table per dimension. The fact revenue is measured in the data warehouse and therefore both dimensions are interconnected to a fact table containing revenue as attribute. The dimensions and the fact table are related by a primary – foreign key relation in which the primary key # of the dimension tables is mapped to the corresponding foreign key $FK\_\ldots$ in the fact table. Figure 3.7 visualizes the data warehouse schema.

For illustration of the open end fuzzy concept, the dimension client with a fuzzy classification age can be used. Every customer is classified into old, middle-aged and young customer. As of today, the youngest registered customer is 14 and the oldest is 50. Therefore, a lower knee point can be chosen at the age of 14 and a higher knee point at the age of 50. These knee points will define that every customer under 14 fully belongs to the class membership attribute young and every customer older than 50 fully belongs to the class membership attribute old. When in the future, a customer with age 62 will be registered, he will fully belong to old customer, even if the fuzzy concept today does not have a target attribute value that is over age 50. Figure 3.8 illustrates the open end fuzzy concept with the classification of the new customer.

Limited Fuzzy Concepts

A second kind of fuzzy concepts are limited fuzzy concepts. In limited fuzzy concepts, values outside the initially defined domain are not considered. The lowest

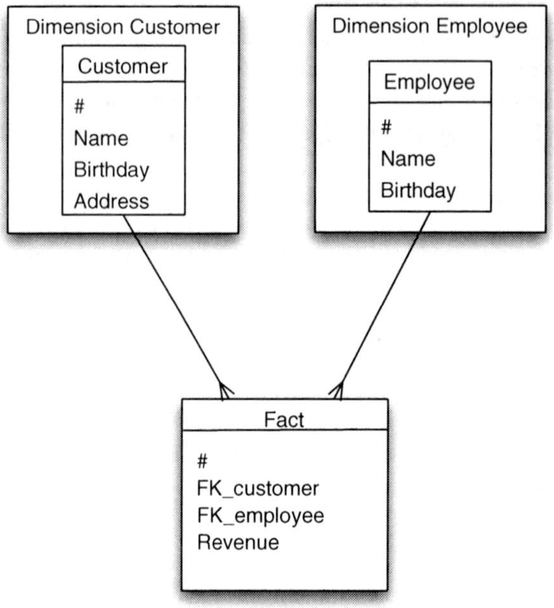

Fig. 3.7 Customer and employee dimension

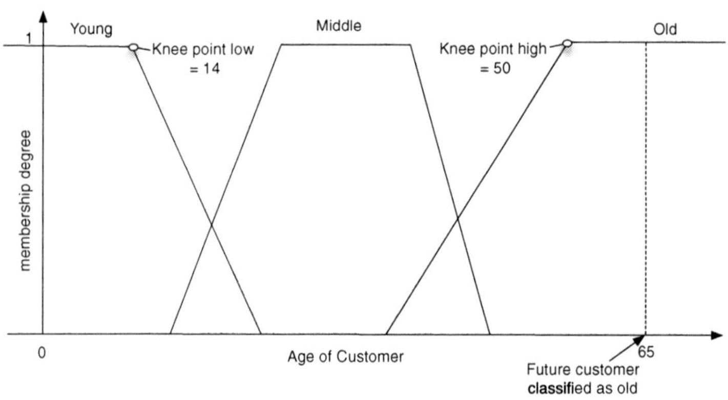

Fig. 3.8 Example of open end fuzzy concept

and highest values of the original value range define the minima and maxima point of the fuzzy concept. Afterwards, all new values out of this range are discarded.

Figure 3.9 show the schematic representation of the limited fuzzy concept. It is notable that the classification of the concept only spans from the minima to the maxima point. All other values of the domain of attributes are not considered for classification.

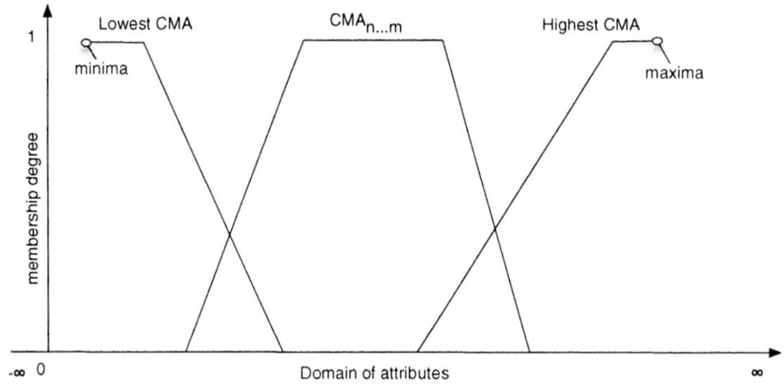

Fig. 3.9 Schematic representation of a limited fuzzy concept

Formally, a limited fuzzy concept can be defined as:

Definition 39 (Limited fuzzy concept). A fuzzy concept for which the value range of the target attribute (TA) is limited with a minima (min) and maxima (max) point; $dom_{TA} = [min, max]$. In this case the domain of the target attributes is not congruent with the complete value range of the dimension attribute or the measure that is classified with this fuzzy concept.

It makes sense to use a limited concept, if the value ranges have static lower and upper boundaries or the fuzzy concept is meant to only classify a subset of the complete value range. The limited fuzzy concept does not consider outliers that are higher or lower than the maxima respectively the minima point. It does not even classify them sharply as in the open end fuzzy concept. In regard to that, the lowest and highest CMA classify more accurately compared to the open end fuzzy concept as they do not considered values that are outliers. The calculation cost of the membership degree is for both concepts almost the same.

As the domain of attributes grows, this fuzzy concept suffers from the same drawbacks as the open end fuzzy concept. In order to maintain a correct classification, the limited fuzzy concept might have to be adapted to the value range if it is frequently fluctuating.

Limited fuzzy concept has lower computation cost than open end fuzzy concepts for the membership degree attribute because it only has to be calculated for values between the minima and maxima point. The fuzzy membership table does not necessarily grow linear to the target attribute table and therefore it might reduce the size of the meta tables structure in the data repository. In contrast, the classification only covers a restricted range. If the value range fluctuates, the limited fuzzy concept can only cover a part of the data. Consequently, it negatively affects the quality of classification.

In Example 10, a fuzzy concept employee age can be applied. In contrast to the customer age, full time employees have a minimum age of 18 and a maximum age of 65. During the summer, students are hired as part time employees for cleaning the

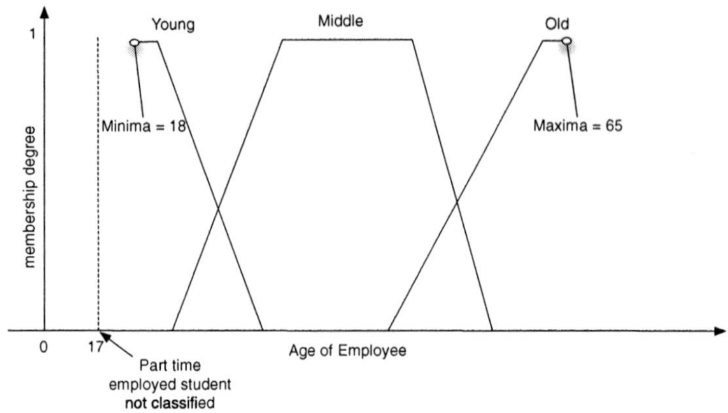

Fig. 3.10 Example of limited fuzzy concept

stock of old movies. These students are often younger than 18 but are out of interest for the classification of employees. Based on this boundaries it is not necessary to capture a domain of attributes containing values below 18 or higher 65 with the fuzzy concept employee age. Therefore a limited fuzzy concept as illustrated in Fig. 3.10 is more suitable than a open end fuzzy concept.

Adaptive Fuzzy Concepts

The adaptive fuzzy concepts uses the complete domain of attributes. This type of fuzzy concept is not defined with fixed points as knee points or minima/maxima points. The CMA are defined relative to the value range and cover a percentile range instead of a defined subset of values. Due to the relative behavior of the adaptive fuzzy concept, it can support a fluctuating set of target attribute values and automatically adapts itself to the new domain of attributes. Therefore, all membership values of all target attribute instances have to be recalculated. This mechanism will always span the fuzzy concept over the actual domain of attributes.

In the schematic representation in Fig. 3.11, the adaptive fuzzy concept always uses the actual domain of attributes as value range. As soon as a new target attribute instance value is inserted, which is outside the actual domain of attributes $D(A)_1$, the membership functions of CMA_1 and CMA_2 are adapted to suit the new domain of attributes $D(A)_2$.

Formally, an adaptive fuzzy concept can be defined as:

Definition 40. A fuzzy concept which has membership functions that define class membership attributes on a percentile range of the domain of attributes of the target attribute.

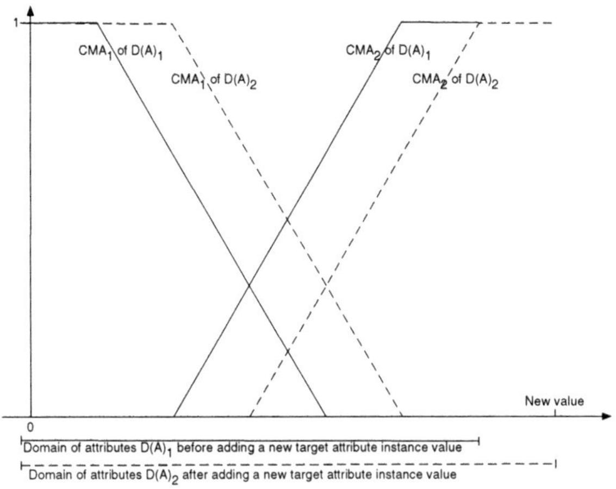

Fig. 3.11 Schematic representation of an adaptive fuzzy concept

Consider Example 10 that uses revenue as its measure in the fact table and applies a fuzzy concept with the class membership attributes high and low to classify revenue. Over time, the company might have periods in which it produces extraordinarily high or low revenue. In order to cover this fluctuation of the domain of attribute revenue, the fuzzy concept can be defined as an adaptive fuzzy concept. The adaptive fuzzy concept checks for every new target attribute value the membership function whether it is within the actual defined domain of attributes or not. In case it is outside the domain of attributes, it determines whether the domain of attributes should be adapted to the value or not. If the value is an outlier it will still get classified based on the old value range and will therefore belong fully to one of the class membership attributes on the outside (low or high). If the domain of attributes is adapted, every membership degree attribute in the membership table, including the new value, will be recalculated based on the new domain of attributes. Figure 3.12 illustrates how the fuzzy concept adapts itself to the new domain of attributes.

Adaptive fuzzy concepts have the disadvantage that they might be sensitive to outlier values. In this context, an outlier can be understood as an exceptional high or low value instance that provokes the recalculation of the fuzzy concept. Therefore, it is necessary to define countermeasures for outlier values in the membership functions. A possible way to define a countermeasure is to calculate the standard deviation and discard values with high variation. For traceability, it might be necessary to archive the old membership degrees. Therefore, given that membership degrees change frequently, it is likely that the adaptive fuzzy concept may lead to a substantial increase in data warehouse size.

Adaptive concepts produces high calculation cost. On every insertion of a target attribute instance, each value must be checked to determine of it is out of the actual domain of attributes and if so, all membership degrees of every instance have to be recalculated. Considering highly distributed data warehouses as in [ISN08]

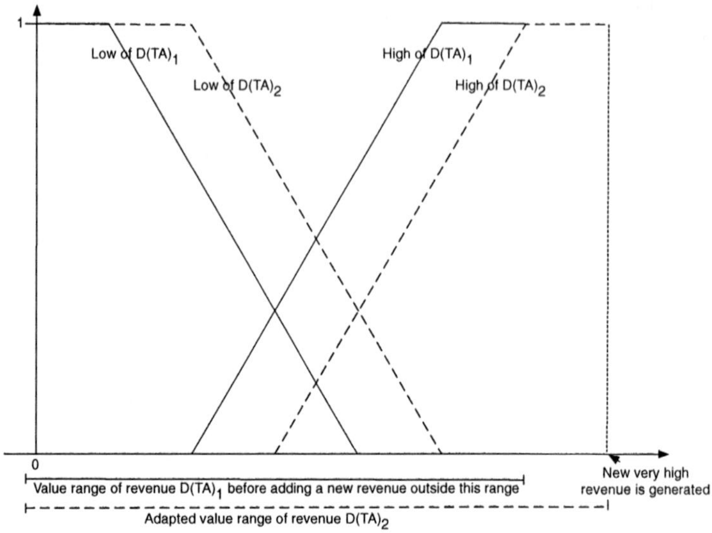

Fig. 3.12 Example of adaptive fuzzy concept

described, the checking mechanisms for the domain of attribute will produce a lot of overhead between the systems and will be a challenging task.

These different types of fuzzy concepts can be combined together. This might be adequate when using big data warehouse systems that have distributed data sources as for example Teradata [Ter11]. If the data is separated by its age in a way that more modern data is in a high available data store and the older data are outsourced on slower data stores it makes sense to integrate adaptive fuzzy concept on the data store with the more recent data. But considering the whole domain of attributes, the adaptive fuzzy concept is seen as a limited fuzzy concept.

3.4.2 Aggregation and Propagation of Fuzzy Concepts

Aggregating of data in a data warehouse directly affects the fuzzy concepts that classify data. Data is grouped together in a more dense view or split to reveal a more detailed view. The grouping is defined by the aggregation function that is often a summation, a product building, a maximization, a minimization, a count or an average building function (see also Sect. 2.1.3 for characteristics of aggregation functions in data warehouse). This aggregation functionality is fundamental to the standard data warehouse operations described in Sect. 2.1.5. In order to be able to classify aggregated data, the fuzzy concepts have to be aggregated too. In the next sections, two methods for aggregation of fuzzy concepts are discussed. The first method aggregates the membership degree instances of the fuzzy concept to a more dense view. The second method is not aggregating membership degree instances, rather it redefines the fuzzy concept with the aggregated data instances as the new target attribute. Therefore, this method is described as propagation of the fuzzy concept as it is propagating the definition of the fuzzy concept and not aggregating values. Each method is illustrated using an example, advantages and drawbacks are depicted.

Aggregation of Fuzzy Concepts

One possibility is to aggregate the membership degree attributes of each target attribute instance into a next higher hierarchy level [Fas09]. Each value instance of the next higher hierarchy level is composed of a set of value instances from the target attribute with respect to summarizability (see Sect. 2.1.3). Consequently, a membership degree attribute for each instance of the next higher hierarchy level can be considered as aggregation of the membership degree attributes of the lower level target attribute instances. In order to illustrate the aggregation of fuzzy concepts, consider the following example:

Example 11. A data warehouse contains a dimension store with two category attributes: store and city. All stores are aggregated to the corresponding city. For all stores their surface is measured and added to the dimension as dimensional attributes on level store. Considering Lehner [Leh98], dimensional attributes can be aggregated on higher hierarchy levels, similarly as it would happen for measures. Therefore, the store surface can be aggregated to the level city.

Further, a fuzzy concept with store surface as target attribute is defined. The fuzzy concept classifies the surface as big, middle and small. Figure 3.13 shows the schema the dimension store and the fuzzy concept. The dimension is set up as two tables store and city. The relation from store to city is maintained by the foreign key relation FK_city on store table to # on city table. The primary key are referenced as #. The fuzzy concept is illustrated as a FCT and a FMT table according to the specification in Sect. 3.2.1.

In Example 11, the average store surface of a city can be calculated by aggregating the surface of all stores in a city with an average function. In order to apply the fuzzy concept store surface on the level city, the membership degree attributes have to be aggregated. By the foreign key relation of stores to cities, it is known which store surfaces are aggregated to a distinct city surface. Subsequently, the membership degree attributes on level store can be identified that aggregate to a membership degree attribute on level city. An additional aggregation function can then be defined that aggregates the membership degree attributes of the stores to the membership degree attribute of the city. In the case of store surface, the arithmetic average of the membership degree attributes of each class membership attribute can be used to generate the corresponding membership degree attributes for the cities. Figure 3.14 shows the aggregation of the fuzzy concept surface from store to city. Store A and store B are aggregated to the city Fribourg. Their surfaces, 20 and 35 m^2

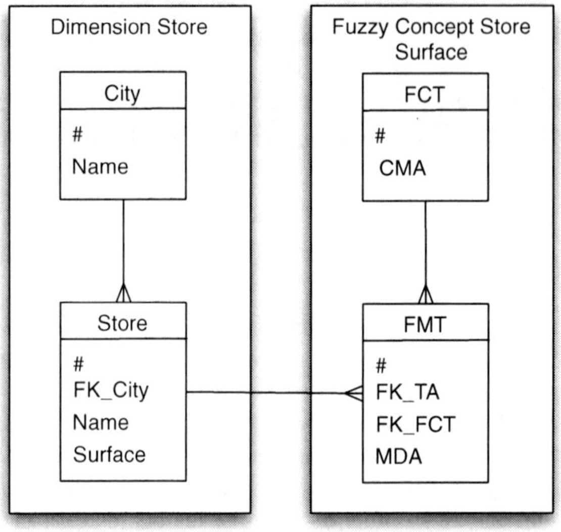

Fig. 3.13 Dimension store with fuzzy concept store surface

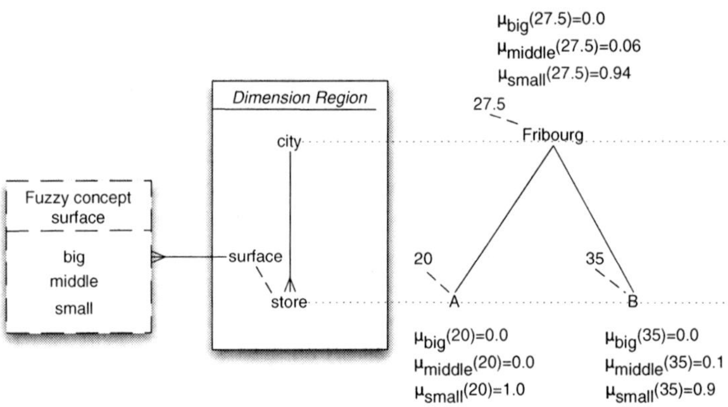

Fig. 3.14 Aggregation of a fuzzy concept

respectively, are aggregated to 27.5 m^2 in city Fribourg using an average function. Both stores have a membership degree represented by μ for the linguistic terms of the fuzzy concept surface in Fig. 3.14. The membership degrees of the city Fribourg are calculated using the arithmetic average of the membership degrees of store A and store B. For example, the membership degree of the linguistic term small for the city Fribourg is calculated as follows: $\mu_{small}(27.5) = \frac{\mu_{small}(20)\times 20 + \mu_{small}(35)\times 35}{55}$. Similarly, the membership degrees for the linguistic terms big and middle can be calculated. Other aggregation function than an arithmetic average may be used for aggregating the membership degrees. Several fuzzy set specific operations have

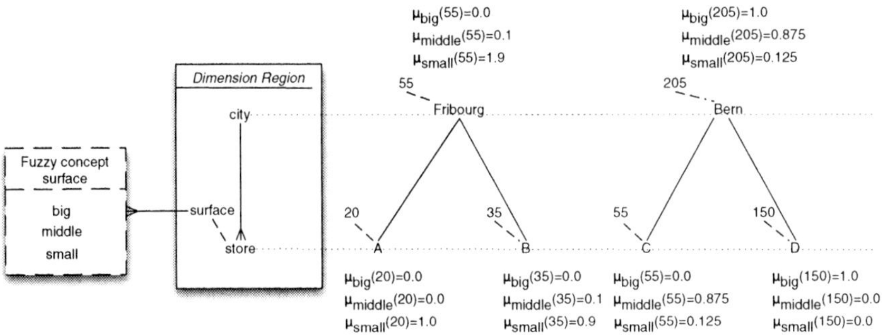

Fig. 3.15 Wrong aggregation of fuzzy concept using summation

been presented in Sect. 2.2.1 which can be used for aggregation of fuzzy concepts. Operations like the *compensatory and* (see Definition 27) might be more adequate for aggregation when the aggregation should better reflect the human perception of the classification.

Aggregation of fuzzy concept may not be semantically correct in all cases. When the city surface is aggregated as the sum of all store surfaces, an aggregation of the membership degree attributes might lead to erroneous classification. If the aggregation of the target attribute is based on a summation function, the domain of the aggregated target attributes does not correspond anymore to the initial domain. Therefore the fuzzy concept itself has to be adapted to the new domain of attributes. This adaption of the fuzzy concept cannot be achieved by aggregating the membership degree attributes. In order to better illustrate the dilemma when the target attribute is aggregated using a summation function, the following two situations are given:

The membership degree attributes are aggregated using a summation function. In this case the city Fribourg would have a store surface of 55 m^2 and it would be classified as $\mu_{small} = 1.0 + 0.9 = 1.9$ and $\mu_{middle} = 0.0 + 0.1 = 0.1$. Figure 3.15 shows this wrong aggregation for the city Fribourg and Bern.

It can be seen that store C in Bern has the same surface as city Fribourg but is classified different. Additionally, the original fuzzy concept had normalized membership degrees in a way that $\mu_{small} + \mu_{middle} + \mu_{big} = 1.0$. The aggregated membership degrees are not normalized anymore and if they would be normalized in a further step, it would appear that the overall surface of city Bern ($\mu_{small\ normalized} = 0.0125/(0.0125+0.875+1.0) = 0.06$) is classified as smaller surface than its child store D ($\mu_{small} = 0.00$).

As a second example, the membership degree attributes are aggregated using the arithmetic average function as described in Fig. 3.14 while the surface is still aggregated using a summation function. In this case, it is possible that a city with a bigger overall surface is belonging more to small than a city with smaller overall surface. Figure 3.16 illustrates that city Basel belongs more to small surface than Fribourg even if the overall surface of Basel is 60 m^2 and Fribourg only has 55 m^2.

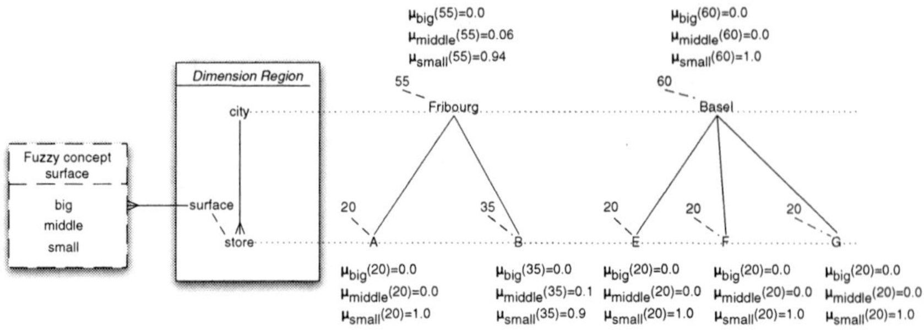

Fig. 3.16 Wrong aggregation of fuzzy concept using average

The reason for this bad classification is the fact that the overall surface of Basel is realized with a higher quantity of smaller stores. Fribourg has fewer stores but they are bigger. The aggregation of the fuzzy concept takes the quantity of the stores into consideration, whereas the summation aggregation of the target attribute does not. Therefore, the classification of the city surface is not consistent anymore.

As a consequence, the aggregation of fuzzy concepts does make sense, if the aggregation of the target attribute is not a summation function. For other aggregation function on target attributes such as minimization, maximization, average and count functions, the aggregation of the fuzzy concept might make sense. Nevertheless, the modeler of the fuzzy concept must always check if the fuzzy concept retains its validity on other hierarchy levels.

Propagation of Fuzzy Concepts

A possible solution to overcome this incorrect classification can be a propagation of the fuzzy concept onto another dimension hierarchy level. In contrast to the aggregation, the propagation of a fuzzy concept does not take the membership degree attributes values into consideration. The linguistic terms and the membership functions are applied to a new hierarchy level as a new variant of the fuzzy concept. The values of the aggregated dimension hierarchy are the target attributes for the new fuzzy concept. The membership degrees are recalculated based on the new domain of attributes. Membership degrees from the fuzzy concept on the lower hierarchy level are not taken into consideration as it would be the case in the aggregation of the fuzzy concept.

For propagating the fuzzy concept store surface from dimension hierarchy level store, a variant of the fuzzy concept is created on the level city. The fuzzy classification table (FCT) is taken from the original fuzzy concept. Whereas, the fuzzy membership table (FMT) has to be newly created for the propagated concept. This is due to the fact that new membership functions are calculating the membership degree based on the new domain of attributes. These newly calculated membership degree attributes are stored in the new FMT. Figure 3.17 illustrates

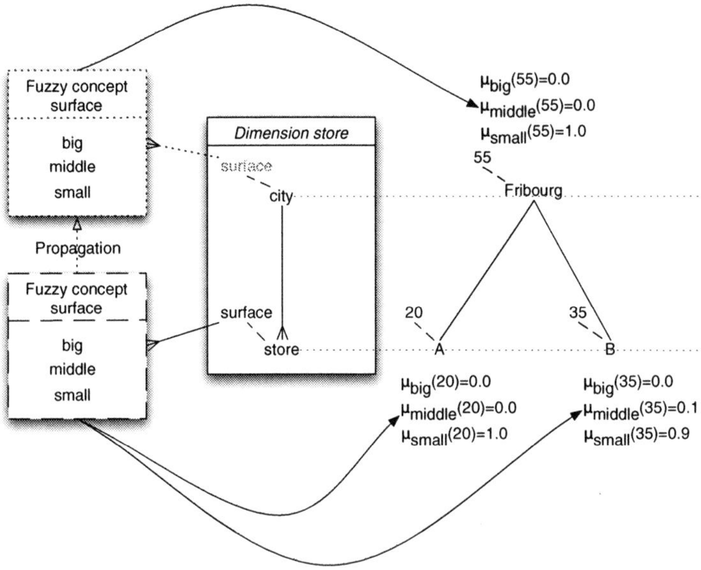

Fig. 3.17 Propagation of fuzzy concept store surface

the propagation of the fuzzy concept store surface from the dimension level store to the level city. The membership degrees of store surface of city Fribourg is now calculated based on the propagated fuzzy concept and with its membership functions. Therefore, these membership degrees are independent from the original fuzzy concept that is calculating only the membership degrees on the store level. The target attribute of the propagated fuzzy concept is the calculated value store surface (in Fig. 3.17 illustrated as grey doted surface attribute on city level) that does not persistently exist in the fuzzy data warehouse. Consequently, the store surface has to be aggregated first at city level in order to calculate its membership degree attributes.

Extending Example 11 with a fact table, another interesting behavior of propagation of fuzzy concepts can be observed:

Example 12. To the dimension store in Example 11 a fact table is added. The fact table contains a measure revenue and the primary key # and the foreign key relation to the store table (FK_store). A second fuzzy concept is added having revenue as the target attribute. Figure 3.18 illustrates the extended example.

Regarding the fuzzy concept revenue on the fact revenue in Example 12, it is possible to propagate and/or aggregate it to the hierarchy level city. For propagating, a variant of the concept including the fuzzy membership table must be defined on city level. The target attribute for this fuzzy concept is the city revenue. It is notable that the city revenue is an aggregated fact value and not persistent in the fuzzy data warehouse (see Sect. 3.4.3). The fuzzy concept is only propagated to city level and not to store level. Propagation allows a modeler of a fuzzy concept to exactly

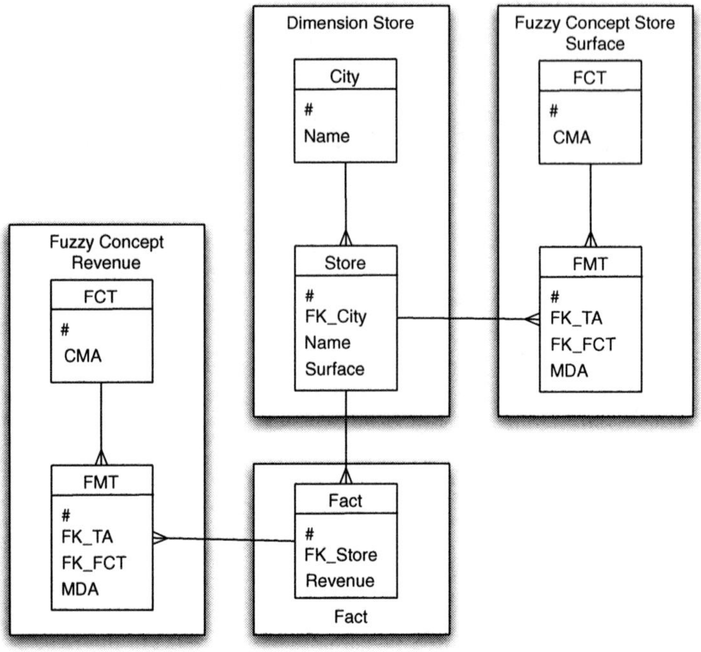

Fig. 3.18 Dimension store and fact revenue with fuzzy concepts

specify on which dimension level the fuzzy concept is propagated. This flexibility is not possible with aggregation as a membership degree attribute on a dimension level is always aggregated from membership degree attributes of the preceding dimension level.

Figure 3.19 demonstrates the propagation of the fuzzy concept revenue to the dimension level city. Store A and B earned multiple revenues of 5 CHF. For every revenue a new instance is stored in the fact table. The total revenue of a city is the sum of all revenues earned by stores. Each revenue has a membership degree for each linguistic term in the fuzzy concept revenue ($\mu_{high}, \mu_{middle}, \mu_{low}$) as shown in Fig. 3.19. All realized revenues belong fully to low revenue in this example. For the city hierarchy level the revenues are aggregated to the city Fribourg and the fuzzy concept is propagated (dashed fuzzy concept city revenue). The city revenue is classified according to the propagated fuzzy concept. Hence, the membership degree attribute on the different hierarchy levels are independent of each other and get defined by the corresponding membership functions.

As for each propagated fuzzy concept, a new fuzzy membership table must be created, it is possible to define different characteristics for propagated fuzzy concept (see also Guideline 3.2 in Sect. 3.2.3). For instance, the base fuzzy concept might be an adaptive concept and the propagated one an open end fuzzy concept.

In technical aspects, propagated fuzzy concepts are only loosely related to its base fuzzy concepts. It may be possible to reuse the fuzzy classification table of

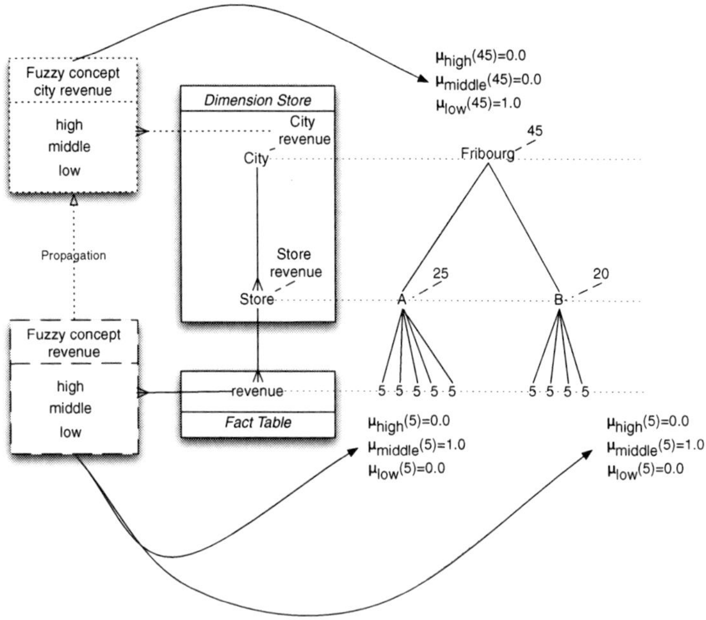

Fig. 3.19 Propagation of a fuzzy concept

the base concept to reduce the amount of extra tables and to limit the resources in the fuzzy data warehouse repository. Besides that, the propagated fuzzy concept is independent of the base concept. In contrast to a propagation, an aggregation of a fuzzy concept creates no variant and the aggregated membership degrees reflect a classification on the base fuzzy concept.

Determining if a fuzzy concept should be propagated, aggregated or both, is a complex task and depends on the nature of the target attribute and the semantic meaning of the classification. In general, the modeler of the fuzzy concept should consider different aspects. The most important aspect is to check if the fuzzy concept retains its semantic correctness on other hierarchy levels. For instance, a limited fuzzy concept might not be propagated if all the new values are outside the defined domain of attributes; respectively the value concept has to be redefined. Wheter the aggregation or the propagation of a concept makes sense is often not based on technical constraints but on the modelers perception of how the fuzzy concept should classify target attributes.

3.4.3 Persistency of Target Attributes

A key feature of data warehouses is the ability to aggregate measures or dimensional attributes over dimensions for analysis purposes. These aggregated values are generally not stored persistently in the data warehouse and will be lost once the analysis is closed. Fuzzy concepts can be defined on measures and dimensional attributes, as described in Sect. 3.2.2, that are persistently stored in the table structures of a data warehouse. An additional challenge is to define fuzzy concepts that have non persistent measures or dimensional attributes as target attributes (briefly addressed in Sect. 3.4.2). The ability to define fuzzy concepts on non persistent target attributes enables propagation of fuzzy concepts on other dimension levels or dimensions (when the fuzzy concept is defined on a fact).

In order to define how fuzzy concepts on non persistent target attributes behave, the propagation of the fuzzy concept revenue in Example 12 is analyzed in greater detail. As described in Fig. 3.19 the fuzzy concept revenue is propagated on dimension level city. First, revenue is aggregated to city revenue. City revenue is a non persistent value as it is calculated just in time of the propagation. Next, the membership degree attributes of the fuzzy concept city revenue are calculated with the city revenue as target attribute. The membership degree attributes are then stored persistently in the corresponding fuzzy membership table of the fuzzy concept city revenue. It must noted that before being able to create the membership degrees, an additional aggregation step has do be done to calculate the volatile target attributes. The key relations from the fuzzy membership table to the target attribute (see Definition 36) still can be realized as the target attribute city revenue belongs distinctly to a city and cities are persistently stored in the city table.

The calculation costs of a fuzzy concept with volatile target attributes depend on the type of fuzzy concept and how often the values, from which the target attribute is aggregated, are changing. Considering the propagated fuzzy concept described in Fig. 3.19, calculation cost might be high because revenue in city Fribourg might be realized regularly and therefore the city revenue is changing frequently. The membership degree attributes have to be recalculated accordingly. When the propagated fuzzy concept is an adaptive fuzzy concept, city revenue for all cities has to be aggregated in order to classify a single city revenue. Consequently, a propagated adaptive fuzzy concept with frequently changed target attribute generates high maintenance costs. In contrast, a propagated non adaptive fuzzy concept with more stable target attributes might have a similar maintenance effort as a fuzzy concept with persistent target attributes. An example for a propagated fuzzy concept with more stable volatile target attributes is the fuzzy concept city surface described in Fig. 3.17. The overall store surface of a city is not as frequently changing as the earned overall revenue in a city. Therefore, it might be worthwhile to use an open end fuzzy concept for the fuzzy concept city surface. When using an open end fuzzy concept, the membership functions do not depend on knowing the complete value range of all target attributes. Because of this, it is not necessary to aggregate the store surfaces of all the cities when calculating the membership degree attributes for a single city. Furthermore, the city store surface is only changing when a store is added or removed in the city. Modifying category attributes in a dimension is less frequent than adding new fact instances (i.e. revenue). Consequently, the membership degree attributes of the fuzzy concept city store surface have to be

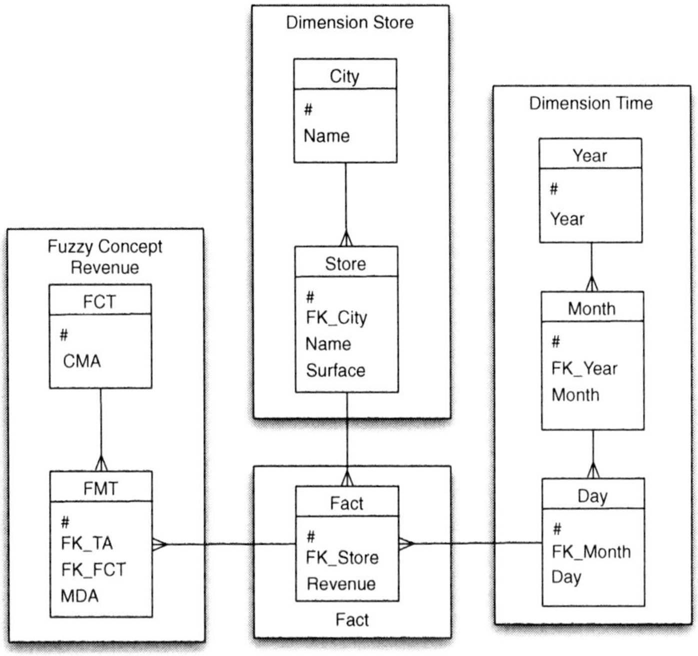

Fig. 3.20 Dimensions time and store, fact revenue and fuzzy concept revenue

adapted less frequently than the membership degree attributes of the fuzzy concept city revenue.

Fuzzy concepts with a dimensional attribute as target attribute are limited to aggregate or propagate over a single dimension. Whereas fuzzy concepts defined on facts might be propagated to more than one dimension. When a dimension time is added to Example 12, the fuzzy concept revenue might be propagated over time in order to receive the fuzzy concept monthly revenue. It should noted that this is a new instance of a propagated fuzzy concept and does not share target attributes with the fuzzy concept city revenue.

In addition, fuzzy concepts based on facts might be propagated over more than one dimension at a time. In Sect. 2.1.5, the classical OLAP operations were presented. Slice and dice operations involve several dimensions at a time in order to analyze a fact. Fuzzy concepts defined on a fact might be propagated on the result set of such an operation. In order to illustrate this, Example 12 is extended with dimension time:

Example 13. To the dimension store and the fact table, a new dimension time is added. This dimension consists of the category attributes day, month and year. The relations between the category attribute is realized with the foreign key attributes FK_month and FK_year in the corresponding tables. For simplicity, the fuzzy concept store surface is removed from the example. Figure 3.20 illustrates the new example.

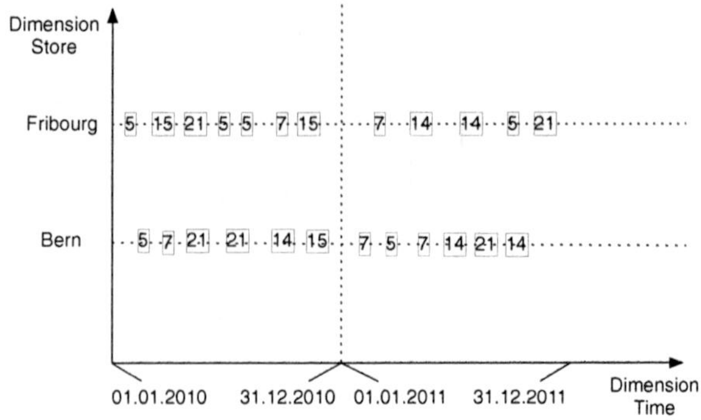

Fig. 3.21 Representation of revenue per city and time

Looking at the cube of the slice operation city revenue in year 2010, a fuzzy concept city revenue 2010 can be propagated on the resulting cube. This fuzzy concept differs from the previously specified fuzzy concept city revenue as the target attribute is the aggregated revenue per city from 2010 and not the overall revenue of the city. Therefore, the classification scope of the fuzzy concept is different. In Fig. 3.21 the X axis of the graph represents the cities of dimension store, the Y axis shows the time. Each city earned different revenues over time, symbolized as boxed figures in the grid. When aggregating the revenue of city Fribourg, the overall revenue is 134 CHF (target attribute for fuzzy concept city revenue) and for the sliced cube the overall revenue for Fribourg is 73 CHF in 2010 (target attribute of the fuzzy concept city revenue in 2010). It is obvious that both aggregation operations lead to different revenues for each city and that the fuzzy concepts therefore classify different scopes.

The application of a fuzzy concept to volatile cubes demands a lot of effort for storing the membership degree attributes in a persistent fuzzy membership table. It would be necessary to create a fuzzy membership table for each possible combination of operations in order to store the appropriate membership degree attributes. Therefore, fuzzy concepts applied to resulting cubes of operations might only be calculated at the time of the analysis and would not be persistently stored at all. Fuzzy concepts with non persistent target attributes can be characterized as volatile fuzzy concepts.

3.4.4 Metaschema for Fuzzy Concepts

At the end of the modeling process of the fuzzy data warehouse (see Sect. 3.3) some additional information about fuzzy concepts, which is not directly obvious in the meta table structure of a fuzzy concept, has to be specified. This additional information contains the characteristics for each fuzzy concept that was specified in the modeling process. Similar to specification of characteristics of facts, the fuzzy concept characteristics can be specified in a meta schema of the data warehouse. The meta schema of data warehouses provides the necessary information how the data warehouse business logic treats a data warehouse element [MT02]. For instance, in the case of facts, the aggregation function is defined. Additionally, how the table structure of the data warehouse is built is defined. With this information, the business logic of the data warehouse can then execute higher level operations, as the typical OLAP operations (Sect. 2.1.5), in a consistent manner.

In order to provide consistent operations on fuzzy concepts, the data warehouse needs to know the nature of a fuzzy concept. The target attribute on which the fuzzy concept is based has to be known, as well as how it is considering the domain of attributes of the target attribute (the type of fuzzy concept). This information tells the data warehouse if the membership degree attributes of the fuzzy concept have to be recalculated after a modification of the target attribute; for instance when using an adaptive fuzzy concept and/or when having volatile target attributes. When using a persistent target attribute, the recalculation takes place when the target attribute is modified. In the case of a volatile target attribute, the recalculation depends on a data warehouse element (base target attribute) from which the volatile target attribute is aggregated. Therefore, the base target attribute must additionally be specified.

One major information about the nature of a fuzzy concept is the membership function. For each class membership attribute of a fuzzy concept, a membership function is defined in order to calculate the corresponding membership degree attribute. The calculation of the membership degree attribute might happen in a business layer or on the database level within a trigger function. The meta schema must contain the information for building these membership functions for each class membership attribute. Section 3.4.5 covers the calculation of membership function including an example in greater detail.

After the nature of a base fuzzy concept is defined in the meta schema, the ability to apply the fuzzy concept to different aggregation levels has to be specified. A fuzzy concept with a dimensional attribute as target attribute might be aggregated and propagated over the dimension. If the target attribute is a fact, the fuzzy concept might be aggregated or propagated over all the dimensions and over resulting cubes of operations in which the fact is involved. Aggregation and propagation do not exclude each other. A fuzzy concept can be applied to different hierarchy levels with both methods. Subsequently, for each method it has to be specified if it is valid and how it is applied. For aggregation, the aggregation function has to be specified and for propagation, the propagated fuzzy concepts have to be specified. When having facts as target attribute, the propagation of the fuzzy concept on a resulting cube from an operation results in a volatile fuzzy concept; a fuzzy concept without any fuzzy membership table or fuzzy classification table. In this case the class membership attributes and the membership functions have to be defined. With this information, the business logic of the data warehouse can calculate the classification during the execution of the operation.

In order to fully specify a fuzzy concept in the meta schema, the following characteristics have to be specified:

- Target attribute of the fuzzy concept
- Persistency of the target attribute (volatile or persistent)
- In case of a volatile target attribute, the base target attribute
- Membership function for each class membership attribute
- For aggregation of the fuzzy concept, the aggregation function
- For propagation of the fuzzy concept, the propagated fuzzy concept
- The volatile fuzzy concept for resulting cubes of operations

3.4.5 Calculation of Membership Function

When using persistent fuzzy concepts, the membership function for calculating the membership degrees can be realized in the database storage system. Therefore, trigger functions can be used for calculating and writing the membership degrees into the fuzzy membership table. The advantage of realizing the membership functions on this low level is the ability to automatically trigger the calculation of the membership degree as soon as a new target attribute instance is added to the fuzzy data warehouse. Handling membership function in business logic of the fuzzy data warehouse results in multiple database connections and more overhead for calculation, and therefore might be slower and less efficient.

Example 14 describes a persistent fuzzy concept and how the different components might be realized in a database storage system. The storage system on which this example is based is the relational database system PostgreSQL [Pos11a]. The trigger functions and SQL statements can differ on other database storage systems. The structure of the functions and statements and differences to other systems are not discussed in the example as it is already illustrated in detail in the PostgreSQL documentation [Pos11b].

Example 14. Figure 3.22 shows two fuzzy classes "low" and "high" and the membership functions μ_{low} and μ_{high}.

Using the classes in Fig. 3.22 the following membership functions are constructed:

$$\mu_{low}(TA) \begin{cases} 1 & \text{for } TA \leq 50 \\ 0 & \text{for } TA \geq 120 \\ 1 - \frac{TA-50}{120-50} & \text{for } 50 < TA < 120 \end{cases}$$

$$\mu_{high}(TA) \begin{cases} 0 & \text{for } TA \leq 80 \\ 1 & \text{for } TA \geq 150 \\ \frac{TA-80}{150-80} & \text{for } 80 < TA < 150 \end{cases}$$

A trigger has to be built that inserts two new data tuples in the fuzzy membership table for every new target attribute. One tuple holds the degree of $\mu_{high}(TA)$, the

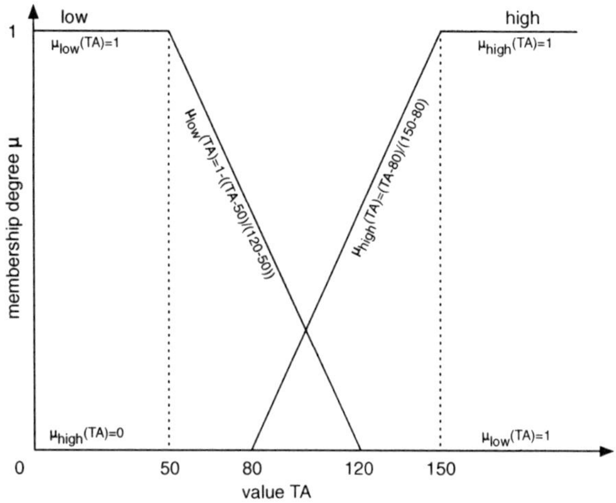

Fig. 3.22 Membership function for two fuzzy classes

other tuple holds the degree $\mu_{low}(TA)$. The code example in Listing 3.2 shows a trigger function calculating and inserting the membership degrees in the fuzzy membership table. Listing 3.1 shows the SQL statements for creating the involved tables in the trigger. The table "Target" is the table in which the target attribute is added, the table "fmt" is the fuzzy membership table and the table "fct" is the fuzzy classification table.

```
— Create the target attribute table
CREATE TABLE Target (
        id as serial primary key,
        ta as integer
        );
— Create the fuzzy classification table
CREATE TABLE fct (
        id as smallint primary key,
        cma as varchar(15)
        );
— Add the class membership attributes to the classification
   table
INSERT INTO fct (id, cma) values
        (1, 'low'),
        (2, 'high');
— Create the fuzzy membership table
CREATE TABLE fmt(
        fct_id as smallint not null,
        ta_id as integer not null,
        mda as numeric(3,2),
        primary key (fct_id, ta_id)
        );
```

Listing 3.1 Table creation statements

```
CREATE FUNCTION fc_memership() RETURNS trigger AS $$
        BEGIN
                -- TA lower 50
                IF NEW.ta < 50 THEN
                        insert into fmt (fct_id, ta_id, mda)
                            values (1, NEW.id, 1.0);
                        insert into fmt (fct_id, ta_id, mda)
                            values (2, NEW.id, 0.0);
                END IF;
                -- TA higher 150
                IF NEW.ta > 150 THEN
                        insert into fmt (fct_id, ta_id, mda)
                            values (1, NEW.id, 0.0);
                        insert into fmt (fct_id, ta_id, mda)
                            values (2, NEW.id, 1.0);
                END IF;
                -- TA between 50 and 80
                IF 50 < NEW.ta < 80 THEN
                        insert into fmt (fct_id, ta_id, mda)
                            values (1, NEW.id, (1-(NEW.TA-50)
                            /70));
                        insert into fmt (fct_id, ta_id, mda)
                            values (2, NEW.id, 0.0);
                END IF;
                -- TA between 80 and 120
                IF 50 < NEW.ta < 80 THEN
                        insert into fmt (fct_id, ta_id, mda)
                            values (1, NEW.id, (1-(NEW.TA-50)
                            /70));
                        insert into fmt (fct_id, ta_id, mda)
                            values (2, NEW.id, (NEW.TA-80)/70));
                END IF;
                -- TA between 120 and 150
                IF 50 < NEW.ta < 80 THEN
                        insert into fmt (fct_id, ta_id, mda)
                            values (1, NEW.id, 0.0);
                        insert into fmt (fct_id, ta_id, mda)
                            values (2, NEW.id, (NEW.TA-80)/70));
                END IF;
                RETURN Null;
    END;
$$ LANGUAGE plpgsql;

CREATE TRIGGER fc_membership After INSERT OR UPDATE ON Target
    FOR EACH ROW EXECUTE PROCEDURE fc_membership();
```

Listing 3.2 PostgreSQL trigger for membership functions

If the fuzzy concept is volatile (see Sect. 3.4.3), the membership function calculation has to be triggered in the business logic as the database storage system does not recognize temporary cubes. But similar to persistent fuzzy concepts, the membership function might be calculated on the database level as a stored procedure. The business logic can then call the stored procedure that creates a temporary cube with the classification of the fuzzy concept applied. In the second step, the classified cube can be returned as a result. The next example describes a fuzzy concept on the fact revenue, which can be applied as a volatile fuzzy concept on non persistent cubes.

Example 15. A fuzzy concept is defined for the fact revenue. The fuzzy concept consists of the linguistic terms high, middle and low. The membership functions do not consider a fixed value range, rather they calculate the membership degrees based on the arithmetic mean. Consequently, the fuzzy concept can be characterized as an adaptive fuzzy concept. The membership functions are as follows:

$$base = 2 * \frac{\sum_{i=1}^{n} TA_i}{n}, \text{ where } n \text{ is the number of target attributes}$$

$$\mu_{low}(TA) \begin{cases} 1 & \text{for } TA \leq 0.2 * base \\ 0 & \text{for } TA > 0.5 * base \\ \frac{0.5*base-TA}{0.5*base-0.2*base} & \text{for } 0.5 * base < TA < 0.2 * base \end{cases}$$

$$\mu_{middle}(TA) \begin{cases} 1 & \text{for } 0.5 * base \leq TA \leq 0.7 * base \\ 0 & \text{for } TA > base \\ \frac{TA-0.2*base}{0.5*base-0.2*base} & \text{for } 0.5 * base < TA < 0.2 * base \\ \frac{base-TA}{base-0.7*base} & \text{for } 0.7 * base < TA < base \end{cases}$$

$$\mu_{high}(TA) \begin{cases} 0 & \text{for } TA \leq 0.7 * base \\ 1 & \text{for } TA \geq base \\ \frac{TA-0.7*base}{base-0.7*base} & \text{for } 0.7 * base < TA < base \end{cases}$$

The fact revenue can be propagated over dimensions using roll-up operations. Using dimensions time and product, it might be possible to analyze the monthly revenue of movies, where movies belonging to the dimension product and month to time. A new cube results from such an analysis that has a new value range of revenue. The described adaptive fuzzy concept can be propagated on the resulting cube using the propagation described in the Sect. 3.4.2. The membership degrees of the volatile fuzzy concept revenue of the resulting cube have to be calculated during the analysis. In order to do this, the SQL statement of the cube might be passed to a stored procedure that creates a temporary table to store the cube enriched with the membership degrees for the revenue of the cube. In the second step, the temporary table can be queried for receiving the fuzzified cube. Listing 3.3 shows an example of such a stored procedure. In order to execute the procedure, the SQL statement of the cube has to be passed as first argument and the attribute in the cube that represents the target attribute as second argument.

```
create or replace function temp.fc_revenue(cube text ,
   ta text)
returns void as $$
declare fields text[];
declare procfield text[][];
declare itr integer;
declare str text;
declare str2 text;
declare cur record;
declare mem numeric;
declare base numeric;
begin
— Parse the cube SQL statement for the column fields
   (select statement) and put them into the field
   array
str := '';
fields := regexp_split_to_array(regexp_replace(
   regexp_replace(cube, E'select', ''), E'from.*', '')
   , ',');
for itr in 1 .. array_upper(fields ,1) loop
        str := str || '{';
        if position(E'"' in fields[itr]) <> 0 then
                str := str || E'"\\"' || substring(
                   fields[itr] from '"(.*)"') || E'
                   \\"",';
        else
                str := str || E'"\\"' || fields[itr
                   ]|| E'\\"",';
        end if;
        — If column is MDA, SUM or AVG it is numeric
           else it is text type
        if position(E'.mda␣' in fields[itr]) <> 0 then
                str := str || '"decimal(3,2)"';
        elsif position(E'␣sum(' in fields[itr]) <> 0
           then
                str := str || '"numeric"';
        elsif position(E'␣avg(' in fields[itr]) <> 0
           then
                str := str || '"numeric"';
        else
                str := str || '"text"';
        end if;
        if itr < array_upper(fields ,1) then
                str := str || '},';
```

```
        else
                str := str || '}';
        end if;
end loop;
-- Create a temporary table with 3 columns for the
   additional MDA's and a primary key
str2 := E'{"\\"mda␣high\\"",␣numeric},{"\\"mda␣middle
   \\"",numeric},{"\\"mda␣low\\"",␣numeric},{"pk","
   serial␣primary␣key"}';
procfield := '{' || str || ',' || str2 || '}';
str2 := 'create␣table␣temp.revvol␣(␣';
for itr in 1 .. array_upper(procfield,1) loop
        str2 := str2 || procfield[itr][1] || '␣' ||
            procfield[itr][2];
    if itr < array_upper(procfield,1) then
                str2 := str2 || ',␣';
        else
                str2 := str2 || ')';
        end if;
end loop;
drop table if exists temp.revvol;
execute str2;
-- Insert the SQL result of cube into the temporary
   table
procfield := '{' || str || '}';
str2 := 'insert␣into␣temp.revvol␣(';
for itr in 1 .. array_upper(procfield,1) loop
        str2 := str2 || procfield[itr][1];
        if itr < array_upper(procfield,1) then
                str2 := str2 || ',␣';
        else
                str2 := str2 || ')␣';
        end if;
end loop;
str2 := str2 || cube;
execute str2;
-- In order to prepare for the fuzzy classification (
   adaptive fuzzy concept) of TA, prepare the base
str := 'select␣avg(␣' || ta || '␣)␣from␣temp.revvol';
execute str into base;
base := 2 * base;
str2 := 'select␣pk,␣' || ta || '␣as␣ta␣from␣temp.
   revvol';
-- Classify TA with the fuzzy concept
```

```
for cur in execute str2 loop
        if cur.ta >= base then
                update temp.revvol set "mda␣high" =
                    1.0, "mda␣middle" = 0.0, "mda␣low"
                    = 0.0 where pk = cur.pk;
        elsif cur.ta >= 0.5 * base and cur.ta <= 0.7
            * base then
                update temp.revvol set "mda␣high" =
                    0.0, "mda␣middle" = 1.0, "mda␣low"
                     = 0.0 where pk = cur.pk;
        elsif cur.ta <= 0.2 * base then
                update temp.revvol set "mda␣high" =
                    0.0, "mda␣middle" = 0.0, "mda␣low"
                     = 1.0 where pk = cur.pk;
        elsif cur.ta > 0.7 * base and cur.ta < base
            then
                mem := (cur.ta -(0.7 * base)) / (base
                    - 0.7 * base);
                update temp.revvol set "mda␣high" =
                    cast(mem as decimal(3,2)) where pk
                    = cur.pk;
                mem := (base - cur.ta) / (base - 0.7 *
                     base);
                update temp.revvol set "mda␣middle" =
                     cast(mem as decimal(3,2)), "mda␣
                    low" = 0.0 where pk = cur.pk;
        elsif cur.ta > 0.2 * base and cur.ta < 0.5 *
            base then
                mem := (cur.ta - 0.2 * base) / (0.5 *
                    base - 0.2 * base);
                update temp.revvol set "mda␣middle" =
                     cast(mem as decimal(3,2)), "mda␣
                    high" = 0.0 where pk = cur.pk;
                mem := (0.5 * base - cur.ta) / (0.5 *
                    base - 0.2 * base);
                update temp.revvol set "mda␣low" =
                    cast(mem as decimal(3,2)) where pk
                    = cur.pk;
        end if;
end loop;
end;
$$ language plpgsql;
```

Listing 3.3 Stored Procedure for the Volatile Fuzzy Concept Revenue

This procedure creates a temporary table revvol that can be queried in the second step. This two step procedure allows the processing of a volatile fuzzy concept on a non-persistent cube. The SQL statement in Listing 3.4 exemplifies how the non-persistent cube of the operation *slice*$(<< time.year, time.month, product.movie >, Revenue, R >, time.year = 2010)$ can be fuzzified using the stored procedure of Listing 3.3.

```
select temp.fc_revenue('select␣month.monthname,␣movie.
    name,␣sum(revenue)␣as␣"Revenue"
from␣fact
join␣(day
join␣(month
join␣year␣on␣month.fk_year␣=␣year.pk_year)␣on␣day.
    fk_month␣=␣month.pk_month␣)␣on␣fact.fk_day␣=␣day.
    pk_day
join␣movie␣on␣fact.fk_movie␣=␣movie.pk_movie
where␣year.year␣=␣2010
group␣by␣month.monthname,␣movie.name',
'"Revenue"');
select * from temp.revvol;
```

Listing 3.4 SQL statement for applying a volatile fuzzy concept on a non-persistent cube

The ability to use stored procedures and triggers to implement membership functions depends on the database technology. In PostgreSQL for instance, it is possible to work with arrays and regular expressions in stored procedures, which is an essential element for parsing the cube SQL statement in Listing 3.3. Not all database technologies provide enough functionalities with their stored procedures and it may be necessary to move the membership functions into the business logic of the fuzzy data warehouse application.

3.5 Operations in Fuzzy Data Warehouse

To provide a functional data warehouse, the fuzzy data warehouse has to support the basic OLAP operations defined in Sect. 2.1.5. In this section, the way that fuzzy concepts can be navigated along dimensions, their impact when aggregating facts and how classical operations have to be adapted in order to handle fuzzy concepts correctly and how multidimensional data cube definition has to be extended with fuzzy concepts is discussed.

During the discussion of the classical OLAP operations in Sect. 2.1.5, a small data warehouse example was created for which the Snowflake scheme is presented in Fig. 2.13. This example is later extended with fuzzy concepts to demonstrate the OLAP operations in a fuzzy data warehouse.

Example 16. A fuzzy concept is added to the fact revenue as target attribute. The fuzzy concept revenue contains three linguistic terms "low", "middle" and "high" revenue. It is characterized as an adaptive fuzzy concept and the membership functions are as follows:

$$R \textit{ is the ordered set of revenue in a way that}$$

$$x_n \in R : R = \{x_1, \ldots, x_{|R|}\} \wedge x_n < x_{n+1}$$

$$rank(x_n) = \frac{n-1}{|R|-1}$$

$$\mu_{high}(x_n) = \begin{cases} 1, & rank(x_n) \geq 0.8 \\ 0, & rank(x_n) \leq 0.6 \\ \frac{rank(x_n)-0.6}{0.8-0.6} \end{cases}$$

$$\mu_{middle}(x_n) = \begin{cases} 0, & rank(x_n) \geq 0.8 \\ 0, & rank(x_n) \geq 0.2 \\ 1, & 0.4 \leq rank(x_n) \leq 0.6 \\ \frac{rank(x_n)-0.2}{0.4-0.2}, & 0.2 \leq rank(x_n) \leq 0.4 \\ \frac{0.8-rank(x_n)}{0.8-0.6} \end{cases}$$

$$\mu_{low}(x_n) = \begin{cases} 0, & rank(x_n) \geq 0.4 \\ 1, & rank(x_n) \leq 0.2 \\ \frac{0.4-rank(x_n)}{0.4-0.2} \end{cases}$$

The dimensions level store of dimension region is extended with a dimensional attribute surface on which another fuzzy concept is added. The fuzzy concept store surface contains the linguistic terms "big", "moderate" and "small" and is a limited fuzzy concept. The minima point is 50 and the maxima point is 200. The membership function are as follows:

$$\mu_{big}(s) = \begin{cases} 1, & s \geq 180 \\ 0, & s \leq 140 \\ \frac{s-140}{40} \end{cases}$$

$$\mu_{moderate}(s) = \begin{cases} 0, & s \geq 180 \\ 0, & s \geq 80 \\ 1, & 120 \leq s \leq 140 \\ \frac{s-80}{40}, & 80 \leq s \leq 120 \\ \frac{180-s}{40} \end{cases}$$

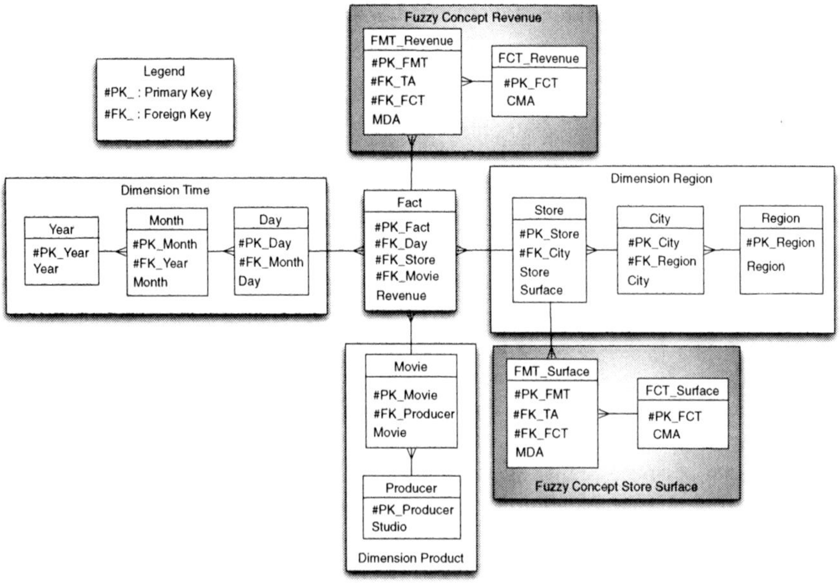

Fig. 3.23 Example fuzzy data warehouse snowflake schema including fuzzy meta tables

$$\mu_{small}(s) = \begin{cases} 0, & s \geq 120 \\ 1, & s \leq 80 \\ \frac{120-m}{40} & \end{cases}$$

$$where \quad s \in dom(region.store.surface)$$

The data warehouse schema presented in Fig. 2.13 can be extended with the fuzzy concept meta table according the method described in Sect. 3.3. The new scheme is presented in Fig. 3.23 including the fuzzy concepts in grey. The dimension tables and the fact table are retained as in Fig. 2.13 with the exception of adding the dimensional attribute surface to table store.

The fuzzy concept revenue is specified to be propagated over all dimensions and over all dimension levels. The propagated fuzzy concept are defined as adaptive fuzzy concept just as the base fuzzy concept. For each dimension level, one fuzzy membership table is created for the propagated fuzzy concepts. The class membership attributes are reused from the base fuzzy concept. Additionally, a volatile fuzzy concept containing the class membership attributes "low", "middle", "high" and a membership function is specified.

The fuzzy concept store surface can only be aggregated over the dimension levels of dimension region. The dimensional attribute surface is aggregated using an average function and therefore an aggregation of the fuzzy concept using the arithmetic average is a valid method for aggregating the fuzzy concept. In this case, no additional tables have to be created.

3.5.1 Classical Data Warehouse Operations in Fuzzy Data Warehouse

Based on the capabilities of aggregating and propagating fuzzy concepts, the classical data warehouse operation can be extended to support fuzzy concept. As mentioned in Sect. 3.4.2, fuzzy concepts can be treated as dimensional attributes. Lehner [Leh98] defines multidimensional objects that are capable of aggregating dimensional attributes and takes them into consideration as segments for slice and dice operations. Consequently, it is possible to aggregate fuzzy concepts and to use them as slicers for slice and dice operations. Agrawal et al. [AGS97] define a model in which a fact is transformed to a dimension and treated accordingly. Based on this possibility, fuzzy concepts on facts can be considered as additional fragments of information behaving just as dimensional attributes on dimensions. As long as a fact can be aggregated over a dimension hierarchy, the fuzzy concept with the fact as target attribute can be aggregated as well. The fuzzy concepts on facts can be considered as segmentation in slice and dice operation just as fuzzy concepts on dimensional attributes. When segmenting a cube with a fuzzy concept, the class membership attributes are the delimiter of a segment. Additionally, the segmentation can be refined using $\alpha - cuts$ on the delimiting class membership attribute. In order to discuss the classical operations, these characteristics of fuzzy concepts have to be taken into consideration. The Definition 10 of a multidimensional cube is adapted accordingly to integrate fuzzy concepts.

Definition 41 (Cube in a fuzzy data warehouse). A cube is as a 4-tuple $< D, F, M, R >$ where

- $D =< D_1, \ldots, D_n, >$ is a list of dimensions, dimensions levels including the dimension attribute or other category attributes separated by a dot.
- $F =< F_1, \ldots, F_j >$ is a list of fuzzy concepts with target attributes that are either facts or dimension attributes or class membership attributes of fuzzy concepts separated by a dot
- a measure M
- R is a set of data tuples $x = \{x_1, \ldots, x_n.f_1, \ldots, f_j, m\}$, where $\forall k \in [1, \ldots, n]$: $x_k \in dom(D_k)$, $\forall l \in [1, \ldots, j] : f_l \in dom(F_l)$ and $m \in dom(M)$ representing the instance values of the cube.

Constraints:

- Every element in D has to be of the same granularity.
- If granularity of $D \leq$ dimension level: $\forall f \in F \, \exists \, d \in D : d = TA_f \vee M = TA_f$
- If $D = \emptyset : \forall f \in F \, \exists \, M = TA_f$
- M is only aggregated over D.

A fuzzy cube in Example 16 can be: << *time.month.month*, *product.producer. studio* >, < *time.month.revenue*, *product.producer.revenue.high* >, *revenue*, *R* >. The fuzzy concept revenue is propagated on dimension time on level month and on dimension product on level producer. The result set of the cube are tuples containing the aggregated revenue grouped by month and product. The membership degree attributes of the propagated fuzzy concepts monthly revenue and producer revenue are added to the data tuples. The fuzzy concepts are not involved in the aggregation of revenue as they provides the same characteristics as dimension attributes in this case. Consequently, the revenue in the result set of the cube << *time.month.month, product.producer.studio* >, <>, *revenue, R* > is exactly the same. This is a direct consequence of the last constraint in Definition 41.

Another interesting fact is that the membership degree attributes originates from fuzzy concepts that are not considering the revenue from the cube. The involved fuzzy concepts are based on revenues that are aggregated on single dimensions; fuzzy concept monthly revenue includes revenue that is only aggregated over dimension time and fuzzy concept producer revenue over dimension product. The fuzzy concept monthly revenue add the attributes *time.month.revenue.high, time.month.revenue.middle* and *time.month.revenue.low* and producer revenue the attribute *product.producer.revenue.high* to the set of *R*. The cube can be extended with a third, volatile fuzzy concept that classifies the actual aggregated revenue: << *time.month.month, product.producer.studio* >, < *time.month.revenue, time.producer.revenue.high, revenue* >, *revenue, R* >. As a result, another three attributes *revenue.high, revenue.middle* and *revenue.low* are added to *R*. Listing 3.5 provides a SQL statement for the cube << *time.month.month, product.producer.studio* >, < *time.month.revenue, product.producer.revenue.high* >, *revenue, R* >. The fuzzy membership tables of the propagated fuzzy concepts monthly revenue and producer revenue are denoted by fmt_revenue_month respectively fmt_revenue_producer in Listing 3.5

```sql
select "month"."month", producer.studio, rev_month_high.
    mda as "time.month.revenue.high", rev_month_middle.
    mda as "time.month.revenue.middle", rev_month_low.
    mda as "time.month.revenue.low", rev_prod.mda as "
    product.producer.revenue.high", sum(fact.revenue) as
     revenue
from fact
join "day" on fact.fk_day = "day".pk_day
join ("month"
join (fmt_revenue_month as rev_month_high
join fct_revenue as fct_high on rev_month_high.fk_fct =
    fct_high.pk_fct and fct_high.cma = 'high') on "month
    ".pk_month = rev_month_high.fk_ta
join (fmt_revenue_month as rev_month_middle
join fct_revenue as fct_middle on rev_month_middle.
    fk_fct = fct_middle.pk_fct and fct_middle.cma = '
    middle') on "month".pk_month = rev_month_middle.
    fk_ta
join (fmt_revenue_month as rev_month_low
join fct_revenue as fct_low on rev_month_low.fk_fct =
    fct_low.fct_id and fct_low.cma = 'low') on "month".
    pk_month = rev_month_low.fk_ta) on "day".fk_month =
    "month".pk_month
```

```
join (movie
join (producer
join (fmt_revenue_producer as rev_prod
join fct_revenue as fct_prod on rev_prod.fk_fct =
    fct_prod.pk_fct and fct_prod.cma = 'high') on
    producer.pk_producer = rev_prod.fk_ta) on producer.
    pk_producer = movie.fk_producer) on fact.fk_movie =
    movie.pk_movie
group by "month"."month", producer.studio, "time.month.
    revenue.high", "time.month.revenue.middle", "time.
    month.revenue.low", "product.producer.revenue.high"
```

Listing 3.5 SQL example for a fuzzy cube

As illustrated in Sect. 3.2.4, the fuzzy data warehouse is an enhanced data warehouse and therefore, a crisp cube (Definition 10) is a fuzzy cube (Definition 41) with $F = \emptyset$. All the classical data warehouse operations (Definitions 11–14) can be applied to a fuzzy data warehouse. Additionally, the classical operations can be extended in order to apply to fuzzy concepts. The following sections discuss these operations in further detail.

Roll-Up in Fuzzy Data Warehouse

A roll-up operation as defined in Definition 11 can be applied to a fuzzy cube. In this case, the resulting cube will have $F = \emptyset$ and all fuzzy concepts of the base cube will be discarded. The following example illustrates a crisp roll-up on a fuzzy cube: $roll-up(< time.month >, < time.month.revenue, revenue >, revenue, R >, f_{time.year}, SUM) =<< time.year >, revenue, R' >$.

In order to include fuzzy concepts in a roll-up operation, it is necessary to know if the fuzzy concepts can be propagated or aggregated over a dimension hierarchy. In case of propagation, additional information is needed on which dimension levels the fuzzy concepts are propagated. Considering these facts, the roll-up operation on a fuzzy data warehouse involving fuzzy concepts can be defined as:

Definition 42 (Roll-up involving fuzzy concepts). $roll-up(C, D', F', f_D, f_F, f_m) = C'$ where

- C is a fuzzy cube (Definition 41)
- D' is the dimension level to which is merged
- f_D is the dimension merge function
- F' is the fuzzy concept which is merged
- f_F is the function defining how the fuzzy concept is merged
- f_m is the measure aggregation function.

The domain (dom_D) of D is a set containing the dimension category attribute instances d. $dom_{D'}(C')$ is calculated by applying the function f_D on the dimension category attribute instances d of $dom_D(C)$:

$$dom_{D'}(C') = \{f_D(d) = d' | d \in dom_D(C)\}.$$

The measure $m_{C'}(d')$ of each instance d' is calculated by applying the function f_m to each $m_C(d)$ in regard to the aggregation function f_D.

$$m_{C'}(d') = f_m(\{t | t = m_C(d) \wedge f_D(d) = d'\}).$$

The domain (dom_F) of F is a set containing the fuzzy concepts f. $dom'_F(C')$ is calculated by applying the function f_F on the fuzzy concepts of $dom_F(C)$:

$$dom'_F(C') = \{f_F(f) = f' | f \in dom_F(C)\}$$

Within a roll-up operation on a propagated fuzzy concept it is possible that $f_F(f) = \emptyset$. Therefore, C' might result in a sharp cube as $dom'_F(C') = \emptyset$.

The function f_F holds the necessary information on how the fuzzy concept can be merged with the next dimension level. f_F must provide at least the following information:

- The type of merging operation: propagation or aggregation
- If aggregation, the aggregation function for F on D
- If propagation, the fuzzy membership table of F'

When F describes a class membership attribute of a fuzzy concept instead of a complete fuzzy concept, the fuzzy concept is rolled-up and the projection on the specific class membership attribute is then applied to the rolled-up cube.

The following operation exemplifies a roll-up of a cube containing the fuzzy concept store surface:

$roll{-}up(<< region.store >, < region.store.surface >, revenue, R >, region.city, region.city.surface, f_D, f_F, SUM) = << region.city >, < region.city.surface >, revenue, R >$

The original cube contains the aggregated revenue of all the stores ordered by the fuzzy concept store surface. The cube resulting from the roll-up operation contains the aggregated revenue of all cities ordered by the fuzzy concept city surface. The fuzzy concept city surface is an aggregated fuzzy concept. This roll-up operation is a binary operation that first aggregates and merges the crisp fact revenue and the crisp dimension region. In the second step, it applies the new fuzzy concept to the data collection. As mentioned in Sect. 3.5.1, the fuzzy concept is not involved in the aggregation of the fact.

Considering F contains a class membership attribute, the fuzzy concept is rolled-up, like in the previous example, within a binary operation. In the third step, the projection on the class membership attribute is applied. To illustrate, the roll-up operation above can be modified in a way that the original cube is $<< region.store >, < region.store.surface.small >, revenue, R >$:

roll–up(<< region.store >, < region.store.surface.small >, revenue, R >, region.city, region.store.city.small, f_D*,* f_F*, SUM) = << region.city >, < region.city.surface.small >, revenue, R >*

When including a class membership attribute, the operation is a trinary operation with the following steps:

- Roll-up of the crisp cube
- Application of the fuzzy concept on the new dimensional level
- Application of the class membership attribute

The next example shows a roll-up operation in which the fuzzy concept store surface might not be defined on dimensional level region. The cube results in a crisp cube as $F' = \emptyset$:

roll – up(<< region.store >, < region.store.surface >, revenue, R >, region.region, region.region.surface, f_D*,* f_F*, SUM) = << region.region >, { }, revenue, R >* where $\nexists$ *region.region.surface.*

Drill-Down in Fuzzy Data Warehouse

As previously stated in Sect. 2.1.5, a drill-down operation is the reciprocal operation of a roll-up. It is not a valid operation if the roll-up operation is not defined in an earlier step. Therefore, a drill-down operation in the fuzzy data warehouse can be defined as:

Definition 43 (Drill-down involving fuzzy concepts). $drill-down(C, D', F', f_D^{-1}, f_F^{-1}, f_m^{-1}) = C'$ where C and C' are cubes according Definition 41. D' is the dimension level to decompose, $C = roll-up(C', D, F, f_D, f_F, f_m)$, F' is the fuzzy concept to decompose and f_D^{-1}, f_F^{-1} resp. f_m^{-1} are the inverse function of f_D, f_F resp. f_m.

In the case of a drill-down operation with a propagated fuzzy concept, it is possible that $F' = \emptyset$. The resulting cube here is a crisp cube.

The following example shows a drill-down operation with a fuzzy cube:

drill-down(C, region.store, region.store.surface.small, $f_D^{-1}, f_F^{-1}, f_m^{-1}$*) =< region.store >, < region.store.surface.small >, revenue, R >*

where C = roll-up(<< region.store >, < region.store.surface.small >, revenue, R >, region.city, region.city.surface.small, f_D*,* f_F*,* f_m *= SUM)*

C' is a cube on dimension level city. It is ordered by the fuzzy concept city surface and limited by the class membership attribute small. The drill-down operation decomposes the cube to dimension level store. The fuzzy concept is not literally decomposed. Rather, the dimension and the fact of the crisp cube are decomposed and the fuzzy concept with a target attribute on the dimension level store is applied to the cube. After this step, the fuzzy concept is limited to the class membership attribute small.

If the fuzzy concept revenue in Example 16 is only propagated to dimension level region and it skips the dimension level city, the following roll-up operation will still result in a fuzzy cube:

$roll\!-\!up(<< region.store >, < region.store.revenue >, revenue, R >, region.region, region.region.revenue, f_{region.region}, f_{region.region.revenue}, f_{revenue} = SUM) = << region.region >, < region.region.revenue >, revenue, R >$

The fuzzy drill-down operation from dimension level region to level city results in a crisp cube as the fuzzy concept revenue is not propagated on level city in this example:

$drill\!-\!down(<< region.region >, < region.region.revenue >, revenue, R >, region.city, region.city.revenue, f^{-1}_{region.region}, f^{-1}_{region.region.revenue}, f^{-1}_{revenue}) = << region.\ city >, \{\}, revenue, R >$

Slice in Fuzzy Data Warehouse

The classical slice operation as defined in Definition 13 extracts a subset of values of a cube that are restricted with a slicer. Applying this to a fuzzy cube, the slicer belongs either to a dimension or to a fuzzy concept. A fuzzy concept as slicer can be restricted either by refining the fuzzy concept with its class membership attributes or by using an $\alpha - cut$ [Zim91] on a class membership attribute. A definition of a slice operation in the fuzzy data warehouse is as follows:

Definition 44 (Slice involving fuzzy concepts). $slice(C, s) = C'$ where C and C' are fuzzy cubes according Definition 41 and s is the element instance that slices cube C in a way that

- $s \in dom(D_m)\ \dot{\vee}$
- $s \in dom(F_m)$ if granularity of F is fuzzy concept $\dot{\vee}$
- $s\ :\ \alpha - cut$ on F if granularity of F is class membership attribute of fuzzy concept.

A slice operation of a fuzzy cube will always result in a fuzzy cube. A slice operation on a crisp cube will also always result in a crisp cube. In order to illustrate, the following operation slices a fuzzy cube according its class membership attributes:

$slice(<< time.month >, < time.month.revenue >, revenue, R >, time.month.revenue.middle) = << time.month >, < time.month.revenue.middle >, revenue, R >$

The slice operation limits the fuzzy concept monthly revenue on its class membership attribute middle. In the resulting cube, the class membership attributes low and high are not considered anymore. A more fine-grained slicing can be achieved when using $\alpha - cuts$ on the class membership attributes. The resulting cube from the slice operation above can be further sliced on the class membership attribute middle with an $\alpha - cut$ of > 0.8:

slice($<< time.month >, < time.month.revenue.middle >, revenue, R >, time.month.revenue.middle > 0.8) = << time.month >, < time.month.revenue.middle >, revenue, R >$

The restriction of the slicer is applied to the class membership attribute middle in a way that only membership degree attributes higher 0.8 are selected in the resulting cube. Multiple $\alpha - cuts$ can be combined in a slice operation. The slice operation below exemplifies a slice selecting all the data tuples that contain a membership degree attribute between 0.6 and 0.8 of middle monthly revenue:

slice($<< time.month >, < time.month.revenue.middle >, revenue, R >, 0.6 \leq time.month.revenue.middle \leq 0.8) = << time.month >, < time.month.revenue.middle >, revenue, R >$

In order to fully include the slice operation of Definition 13, the slice operation on a fuzzy cube can use a dimension element as slicer. The resulting cube of the last slice operation can therefore be restricted to the instance February of dimension category month:

slice($<< time.month >, < time.month.revenue.middle >, revenue, R >, time.month.month =$ "February"$) = << time.month >, < time.month.revenue.middle >, revenue, R >$

Dice in Fuzzy Data Warehouse

A dice operation in a fuzzy cube, just as a slice operation, can restrict fuzzy concepts and dimensions. Furthermore, restrictions of fuzzy concepts and dimension can be combined. The slice operation in Definition 44 can be extended to a dice definition by including logical operators to combine multiple slicers:

Definition 45 (Dice involving fuzzy concepts). $dice(C, \{s_m, \ldots, s_k\}, \{f_m, \ldots, f_{k-1}\}) = C'$ where C and C' are fuzzy cubes according to Definition 41, $s_m, \ldots, s_k$ are element instances that slices cube C in a way that $\forall n \in \{m, \ldots, k\}$:

- $s_n \in dom(D_n) \dot{\vee}$
- $s_n \in dom(F_n)$ if granularity of F_n is fuzzy concept $\dot{\vee}$
- $s_n : \alpha - cut$ on F_n if granularity of F_n is a class membership attribute of fuzzy concept.

and $\forall x \in m, \ldots, k-1 : f_x \in \{AND, OR, NOT\}$ are the logical operators that combine the slicers in a way that f_x combines s_x with s_{x+1}.

A dice operation on a fuzzy cube with two fuzzy concepts is illustrated as follows:

dice(<< time.month, region.city >, < time.month.revenue.middle, region.city.surface >, revenue, R >, {time.month.revenue.middle ≥ 0.7, region.city.surface.big}, {OR}) = << time.month, region.city >, < time.month.revenue.middle, region.city.surface.big >, revenue, R>

The data set of the resulting cube only considers the class membership attribute big of the fuzzy concept city store surface and all the data tuples with a membership degree attribute bigger than 0.7 for middle monthly revenue. A combination of a dimension slicer and a fuzzy concept slicer is shown in the following example:

dice(<< time.month, region.city >, < time.month.revenue.middle, region.city.surface >, revenue, R >, {time.month.revenue.middle ≥ 0.7, time.month.month = "February"}, {AND}) = << time.month,region.city >, < time.month.revenue.middle, region.city.surface.big >, revenue, R >

The resulting cube shows only the data tuples of the month "February" that belong to at least 0.7 to middle monthly revenue.

For the slice and the dice operation, the order how the fuzzy concept or the dimension slicers are applied does not influence the result of the operation. This is due to the fact that no aggregation function is applied to a fact or a dimensional attribute. Therefore, unlike roll-up and drill-down, slice and dice operation are not partitioned in multiple sub operations and can be considered as unary operations.

3.5.2 *Fuzzifying and Defuzzifying Cubes*

Two additional operations can be applied to a cube in the fuzzy data warehouse. Removing the fuzzy concepts of a fuzzy cube will transform it into a crisp cube. Vice versa, if fuzzy concepts are applied to a crisp cube, it transforms into a fuzzy cube. Definition 41 defines a fuzzy cube as a cube having $F = \{F_1, \ldots, F_n\}$ and consequently a crisp cube as a cube having $F = \emptyset$. Fuzzifying a crisp cube thus implies changing the empty set F into a non-empty set. On the contrary, defuzzifying a fuzzy cube removes all fuzzy concept in F in order to have an empty set. The two operations are defined as follows:

Definition 46 (Fuzzifying a cube). $fuzzify(C, fc) = C'$ where $C =< D = \{D_1, \ldots, D_n\}, F = \emptyset, M, R >, fc = \{f_1, \ldots, f_m\}$ are the fuzzy concepts to add in F in a way that $\forall x \in \{1, \ldots, m\} \ \exists \ f_x : TA_{f_x} = dom(D_y) \vee TA_{f_x} = M$

where $y \in \{1, \ldots, n\}$ and the resulting cube is $C' =< D = \{D_1, \ldots, D_n\}, F = \{f_1, \ldots, f_m\}, M, R >$.

It is important to note that only fuzzy concepts can be added to a crisp cube which has target attributes belonging to a dimension or fact of the cube. Consequently, for propagated fuzzy concepts, a derived fuzzy concept with the target attribute belonging to the crisp cube must be defined prior to the fuzzification operation. For an aggregated fuzzy concept, the corresponding membership degree attributes have to be aggregated on the level of the crisp cube.

The contrary operation is defuzzifying: the removal of fuzzy concepts from a fuzzy cube. Definition 47 defines therefore the operation defuzzifying.

Definition 47 (Defuzzifying a cube). $\mathit{defuzzify}(C) = C'$ where C is a fuzzy cube according Definition 41 and $C' =< D = \{D_1, \ldots, D_n\}, F = \emptyset, M, R >$.

Consider the crisp cube $C =<< \mathit{time.day} >, \mathit{revenue}, R >$. When applying the fuzzy concept revenue on level day, it can be transformed to a fuzzy cube as follows:

$\mathit{fuzzify}\,(C, \mathit{time.day.revenue}) = << \mathit{time.day} >, < \mathit{time.day.revenue} >, \mathit{revenue}, \mathrm{R} >= C'$

In a further step, a roll-up operation might be applied to the fuzzy cube. The operation merges from dimension level day to level month and the fuzzy concept is applied accordingly.

$\mathit{roll{-}up}\ (C', \mathit{time.month}, \mathit{time.day.revenue}, f_{\mathit{time.month}}, f_{\mathit{time.month.revenue}}, \mathit{SUM}) = << \mathit{time.month} >, < \mathit{time.month.revenue} >, \mathit{revenue}, R >= C''$

As explained in Sect. 3.5.1, the roll-up operation in this case is a binary operation. First, the dimension level day is merged to month and the daily revenue is aggregated to monthly revenue. In the second step, the fuzzy concept monthly revenue is applied to the cube. The resulting fuzzy cube C'' can now be defuzzified to a crisp cube as follows:

$$\mathit{defuzzify}(C'') =<< \mathit{time.month} >, \mathit{revenue}, R >= C'''$$

The resulting cube C''' is exactly the same cube that would result with a crisp roll-up operation from dimension level day to month as follows:

$$\mathit{roll} - \mathit{up}(<< \mathit{time.day} >, \mathit{revenue}, R >, f_{\mathit{time.month}}, \mathit{SUM}) = C'''$$

This behavior is due to the fact that the fuzzy concepts are meta information in the fuzzy data warehouse and never directly affect the raw data in the original crisp data warehouse. The fact that fuzzy concepts are meta information is also the reason why the inclusion of fuzzy concept is always performed in additional steps where aggregation operations of facts are executed first.

3.5.3 Aggregations with Fuzzy Concept

All the operations in Sect. 3.5.1 consider fuzzy concepts similar to dimension attributes even if the fuzzy concept is applied to facts. Subsequently, the fuzzy concept is never considered in the aggregation function of the fact. The operations are often binary or trinary in order to aggregate first the corresponding fact according to the selectors and then to enrich the result set with the fuzzy concept. The selector is the part of a data set that defines the aggregation and grouping of the fact. In a classical data warehouse, the selectors of a cube are the category attributes of the dimensions. The following cube has the category attribute month as selector and the revenue is aggregated and grouped over the category attribute:

$$C = << time.month >, < time.month.revenue >, revenue, R >$$

The fuzzy concept monthly revenue adds information about the classification of the revenue. In some cases, it might be interesting to analyze the data based on the fuzzy concept as a selector. The fuzzy concepts are then defining directly how the fact is aggregated and grouped. In the example above, when the revenue should be analyzed based on its distribution on the class membership attributes of the fuzzy concept monthly revenue, it acts as an additional selector.

In order to apply a fuzzy concept as selector, the membership degree attributes define how the fact instances are distributed on the different class membership attributes. Hence, the membership degree attributes are a kind of weighting factors to distribute the fact instances. As depicted in Sect. 3.1.2, Delgado et al. in [DMS$^+$04], Scheperle et al. in [SMH04] and Sapir et al. in [SSR08] propose methods for weighting facts with membership degrees during aggregation. Sapir et al. use the concept of bridging tables that is known in classical data warehouses for handling multivalued dimensions.

Bridge tables have been defined by Kimball [Kim98]. A multivalued dimension is a dimension in which an instance of a category attribute might be merged into multiple instances of the next higher category attribute. For that case, a bridge table contains a weighting factor for each possible merging. The weighting factor helps for consistent aggregation operation over the dimension. The Example 17 depicts a multivalued dimension with a bridge table in detail.

Example 17. The dimension producer has the category attribute movie producer. The hierarchy path defines that movies are merged into producer. A movie can be produced as a joint project of several production studios. For instance, the movie “Machete” is produced by the studios “Overnight Films”, “Troublemaker Studios”, “Dune Entertainment III” and “Dune Entertainment”. When the revenue of the movie is aggregated to level producer, the movie revenue must be split and distributed to the different producer studios. Otherwise, the producer revenue would be four times the movie revenue and the aggregation would no longer be consistent. For splitting and distributing the movie revenue a bridge table is added between the two dimension levels. The bridge table contains the relation between

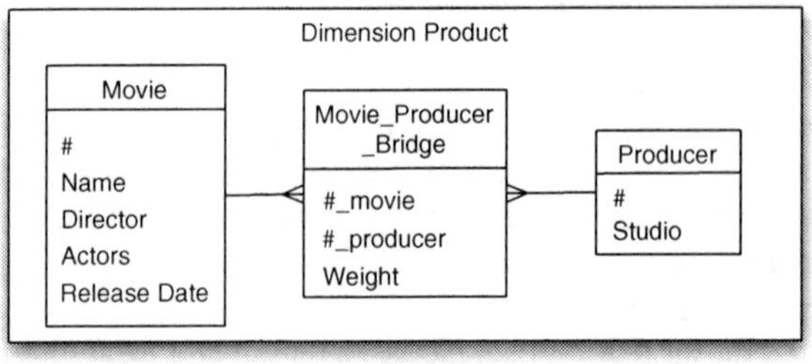

Fig. 3.24 Multivalued dimension with a bridge table

the movie and the producer and the weighting factor that defines how the revenue is distributed between the different studios. When the revenue is distributed equally between the studios, the weighting factor might simply be the number of studios producing the film. The producer revenue of a studio is then calculated by dividing the movie revenue by the weighting factor. Figure 3.24 demonstrates the multivalued dimension product.

Examining bridge tables shows their similar appearance to fuzzy membership tables. Both types of table relate two other tables in a complex to complex relationship (multiple target attributes are related to multiple class membership attributes) and contain an attribute describing a degree of the relation (membership degree or weighting factor). A fuzzy concept can be considered as an additional hierarchy path of the dimension and it can therefore be used as an selector as described by Sapir et al. in [SSR08]. Consider an analysis on the above cube with the fuzzy concept monthly revenue as the selector. The measure has to be partitioned by the membership degree attributes of the fuzzy concept. Therefore the cube is extended as follows:

$$C = << time.month >, < time.month.revenue >, revenue \times time.month.revenue.mda, R >$$

The result set in this case provides a single data row with the corresponding revenue for each class membership attribute and month combination. Example 18 illustrates the differences between the two cubes above based on a numerical example.

Example 18. In a fuzzy data warehouse revenue of 2 month is stored. The dimension time has the category attributes day and month. Day is merged into month. A fuzzy concept on revenue with the class membership attributes low and high are defined. This fuzzy concept is propagated to day and month. When rolling up the revenue on month, the monthly revenue for month January is 500 and the revenue for February is 720. The corresponding membership degree attributes for January high is 0.3, low is 0.7, for February high is 0.5 and low 0.5.

Table 3.3 Result set of cube $C =<< time.month >, <time.month.revenue>, revenue, R >$

Month	CMA high	CMA low	Revenue
January	0.3	0.7	500
February	0.5	0.5	750

Table 3.4 Result set of cube $C =<< time.month >, <time.month.revenue>, revenue \times time.month.revenue.mda, R >$

Month	CMA	Revenue
January	High	150
January	Low	350
February	High	375
February	Low	375

Considering the first cube $C =<< time.month >, < time.month.revenue >, revenue, R >$ the result set contains the month, the membership degree attribute for class membership attribute high, the membership degree attribute for class membership attribute low and the revenue of the month. Table 3.3 represents the result set.

The second cube $C =<< time.month >, < time.month.revenue >, revenue \times time.month.revenue.mda, R >$ uses the membership degrees for partitioning the revenue. Therefore, four data rows will appear in the result set, for each class membership attribute and month combination a single data row. Table 3.4 shows the corresponding result set.

The comparison of the SQL statements to retrieve these cubes shows that in the first cube (Listing 3.6) the class membership attributes are restricting the join clauses and in the second cube (Listing 3.7) the membership degree attributes are involved in the aggregation function of the fact revenue. Both SQL statements are based on the table structure described in Fig. 3.23.

```
select "month"."month", rev_month_high.mda as "time.
    month.revenue.high", rev_month_low.mda as "time.
    month.revenue.low", sum(fact.revenue)
from fact
join "day" on fact.fk_day = "day".pk_day
join ("month"
join (fmt_revenue_month as rev_month_high
join fct_revenue as fct_high on rev_month_high.fk_fct =
      fct_high.pk_fct and fct_high.cma = 'high') on "
    month".pk_month = rev_month_high.fk_ta
join (fmt_revenue_month as rev_month_low
join fct_revenue as fct_low on rev_month_low.fk_fct =
    fct_low.pk_fct and fct_low.cma = 'low') on "month".
    pk_month = rev_month_low.fk_ta) on "day".fk_month =
    "month".pk_month
group by month.month, "time.month.revenue.high", "time.
    month.revenue.low"
```

Listing 3.6 SQL statement for a cube with fuzzy concept as information

```
select "month"."month",fct_revenue.cma, sum(fact.revenue
    * rev_month.mda)
from fact
join "day" on fact.fk_day = day.pk_day
join ("month"
join (fmt_revenue_month as rev_month
join fct_revenue on rev_month.fk_fct = fct_revenue.
    pk_fct) on "month".pk_month = rev_month.fk_ta) on "
    day".fk_month = "month".pk_month
group by "month"."month", fct_revenue.cma
```

Listing 3.7 SQL statement for cube with fuzzy concept as selector

The aggregations in both cube operations are consistent, with respect to summarizability, and therefore the overall revenue of both cubes is equal. It has to be noted that the second cube operation is consistent because the membership degree attributes are normalized and therefore the overall value of all membership degree attributes of a target attribute is 1.0. This condition does not have to be fulfilled for all fuzzy concepts and consequently, the aggregation involving the membership degree attributes might not end in consistent results. In order to normalize fuzzy concepts, each membership degree attribute of a target attribute instance can be divided by the sum of all membership degree attributes [Sch98]. Schepperle et al. describes in [SMH04] extensions to summarizability for aggregation with fuzzy concepts and the main principle is the normalization of the membership degrees. The normalization of the fuzzy concept can therefore be considered as a crucial constraint for consistent aggregation involving fuzzy concept in the aggregation function. If this condition is not met, an additional step that normalizes the membership degree attributes has to be included prior to aggregation of the fact in the cube operation.

Using this functionality of aggregating facts with fuzzy concepts opens up the possibility of combining fuzzy concepts with facts independent of each other. For instance, a fuzzy concept on store surface can be used for the aggregation of the fact revenue. Example 19 below shows how to combine a fuzzy concept of the fact user rating with the fact revenue.

Example 19. A fuzzy data warehouse has two facts "user rating" and "revenue" and the dimensions customer, time and movie. A customer can rent a movie and this generates rental revenue. When returning the movie, the customer rates the movie between 1 and 10. Two fuzzy concepts are defined: the fuzzy concept rating classifies the user rating into good, bad and average, the fuzzy concept revenue classifies the revenue into good, middle and bad. The fuzzy concept user rating can be propagated over dimension customer and over dimension movie. The fuzzy concept revenue can also be propagated over dimension time. Propagation is applied to all category attributes of the dimensions. Figure 3.25 shows the corresponding fuzzy data warehouse schema.

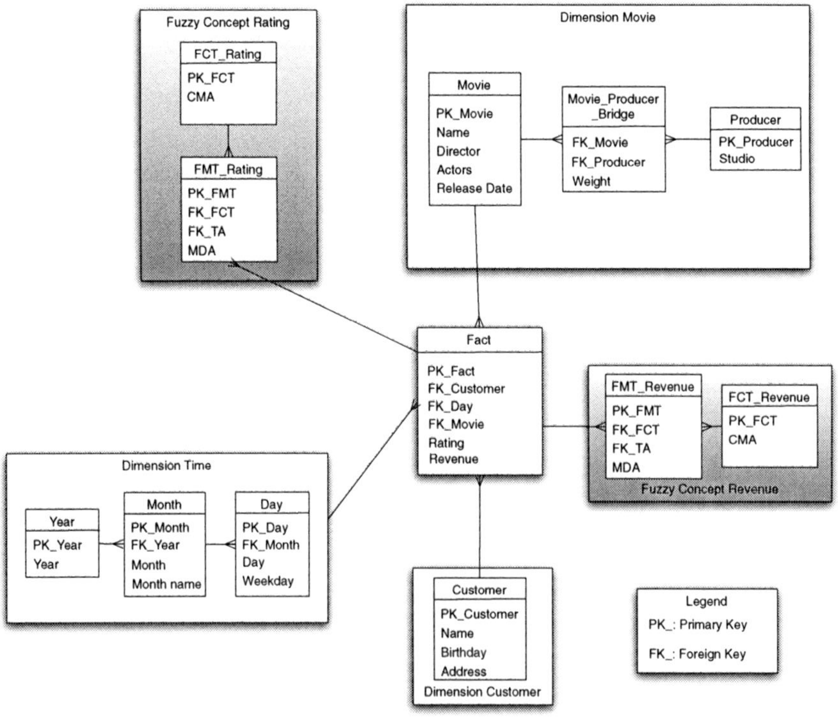

Fig. 3.25 Fuzzy data warehouse schema with dimension customer, movie, time, with facts user rating and revenue

On this fuzzy data warehouse, a cube can be created that shows the monthly revenue for all customers. Additionally, the cube can be extended with the fuzzy concept rating as this concept is propagated on the category attribute customer. The fuzzy cube will therefore be:

$C =<< time.month,\ customer.customer >, < customer.customer.rating >, revenue,\ R >$

It is notable that the fuzzy concept rating is not dependent on the fact revenue. Considering that the fuzzy concept rating is normalized, it can be used for the aggregation of the fact revenue. The new fuzzy cube will be:

$C' =<< time.month,\ customer.customer >, < customer.customer.rating >, revenue \times customer.rating.customer.mda,\ R >$

Cube C' shows the revenue earned grouped by the rating of users and month. Further, the revenue in cube C' can be grouped by the fuzzy concept monthly revenue. The cube C'' will become:

$C'' = << time.month, customer.customer >, < customer.customer.rating, time.month.revenue >, revenue \times customer.customer.rating.mda \times time.month.revenue.mda, R >$

In cube C'' the revenue is aggregated with two fuzzy concepts. Even if the fuzzy concept monthly revenue has a monthly revenue as target attribute, the revenue in C'' is based on month and customer and therefore different from the monthly revenue.

The possibility to combine fuzzy concepts with different target attributes in a fuzzy cube and in the aggregation process allows a more fine grained interpretation of data. In order to achieve a similar analysis like in cube C'' within a classical data warehouse, at least four of the following operations have to be combined:

- Querying cube $C = << time.month, customer.customer >, user\ rating, R >$
- Querying $C' =<< time.month, customer.customer >, revenue, R >$
- Ordering cube C and C'
- Partitioning cube C and C' which is a crisp classification
- Applying the partitioned cube C' to the partitioned cube C by joining over month and customer
- Grouping the joined result set

In database systems as PostgreSQL, it is possible to partition result sets using window functions [Pos11d]. These operations are often limited and do not allow the fuzzy partition of data. The integration of linguistic concepts is completely missing in the steps of a crisp cube explained above. Integration of linguistic concepts would subsequently need further data processing steps. If fuzzy partition and integration of linguistic concepts in a classical data warehouse system is not possible with built-in functions, it must be outsourced to a third party system that is specialized in ranking and classifying data collections. In any case, the classical data warehouse does not provide this seamless integration of fuzzy concepts. The fuzzy data warehouse provides new methods of data interpretation that is closer to natural human language, which is accessible without third party systems.

Part II
Application

Chapter 4
Application of Fuzzy Data Warehouse

This chapter presents a case study in performance measurement using a movie rental company. The case study aims to point out the advantages of a fuzzy data warehouse when classifying elements in a data warehouse. First, the movie rental company and its initial, classical data warehouse are presented in Sect. 4.1. In Sect. 4.2, fuzzy concepts are applied to the data warehouse in order to build a fuzzy data warehouse. Section 4.3 discusses how fuzzy concepts improve analysis in the fuzzy data warehouse.

4.1 The Movie Rental Company

The case study discusses a fictional movie rental company based in Switzerland since 2005. The company has 29 stores spread across 20 cities. It currently offers a collection of 661 movies[1] for rent. All stores are equipped with distribution machines on the outside of the stores where customers can rent the movies outside of the opening hours. For each rental transaction an employee is registered in the data warehouse. If a movie is rent from a distribution machine the employee who was working during the office hours that day is registered. Therefore, to each record of a rental transaction, an employee is listed. The movie rental company has a total of 50 employees. New movies are rent for 7 CHF a day and movies older than 1 year are rent for 5 CHF a day. Each customer is asked to rate the movie when he returns it in order to provide movie charts as a service to the customers. If the customer does not rate the movie, a standard rating value of 5.0 is assumed.

To measure the performance of their business, the movie rental company has implemented a data warehouse. An important fact that the company uses to evaluate its performance is the revenue of every single rental transaction. For each

[1]The data about the movies for this case study was retrieved from the open movie database project using their open API [TMD10].

D. Fasel, *Fuzzy Data Warehousing for Performance Measurement*, Fuzzy Management Methods, DOI 10.1007/978-3-319-04226-8_4,

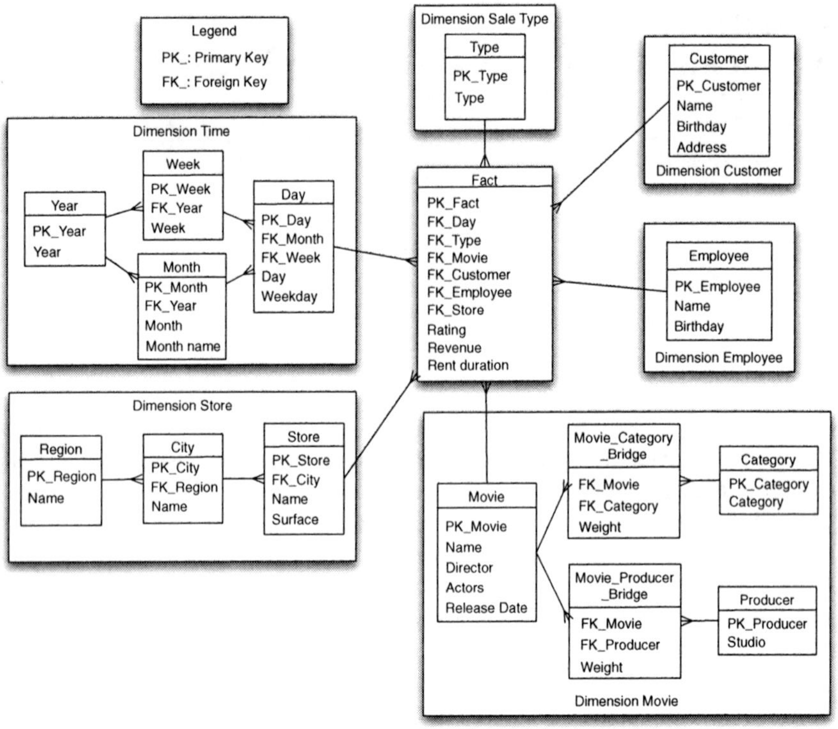

Fig. 4.1 Snowflake scheme of the data warehouse

transaction, information about customer, movie, employee and type is captured. The type dimension classifies each movie as old or new. The movie dimension aggregates movies into two hierarchy paths. One path aggregates producer and the other the category of the movie. Both hierarchy paths are multi-valued hierarchies. A movie can be produced by multiple studios and can be categorized into multiple genres. In order to overcome this difficulty, the concept of bridge tables [KC04, Kim98] (see Example 17 for more details about bridge tables) is used. Additional to the measure revenue, the user rating and rental duration are kept in the fact table. Figure 4.1 presents the snowflake scheme of the movie rental data warehouse.

By using the classical data warehouse shown in Fig. 4.1, the movie rental company can analyze its performance in different aspects. Some examples are given below.

Example 20. The company might be interested in the overall revenues generated by new movies across different regions during 2010. This is a dice operation over a sub cube visualizing the sum of the revenues grouped by regions. The slicers of the dice operation are executed on the dimension time (year = 2010) and the dimension type (type = "new"). Formally, the operation is:

Table 4.1 Result set of Example 20

Region	Overall revenue
German	2, 038, 001.00
French	1, 779, 792.00
Rhaeto-Romanic	141, 652.00
Italian	138, 964.00

$dice(<<region.region, type, year>, revenue, R>, \{type.type = \text{“}new\text{''}, year.year = 2010\}, \{AND\})$.

The corresponding SQL statement is in Listing 4.1.

```
select "region"."region", to_char(sum(revenue), '9G999G999D99')
    as "overall␣revenue"
from fact
join ("day" join ( "month" join "year" on "month".fk_year = "
    year".pk_year and "year".year = 2010) on "day".fk_month = "
    month".pk_month) on fact.fk_day = "day".pk_day
join type on fact.fk_type = type.pk_type and type = 'new'
join (store join ( city join region on city.fk_region = region.
    pk_region) on store.fk_city= city.pk_city) on fact.fk_store
    = store.pk_store
group by region.region
order by "overall␣revenue" desc;
```

Listing 4.1 SQL statement for selecting the revenue per region

The result of the operation is shown in Table 4.1.

Example 21. The movie rental company would like to analyze the revenue sorted by movie producers. Movies from the 20 producers with the highest revenue are preferable for the company when choosing new movies for the portfolio. The revenues of movie producers should be divided with the weighting factor in the bridge table to give meaningful results. A significant amount of revenue comes from films that do not have an assigned production studio. These revenues are lumped into a virtual producer named “Not Known” and should not be considered in the result set. This query can be executed by using a roll-up operation as follows:

$roll-up(<<producer.movie>, revenue, R>, producer.studio, f_{producer.studio}, SUM)$.

The corresponding SQL statement is shown in Listing 4.2.

```
select producer.studio as studio , to_char(round(sum(cast(revenue
     as decimal)/mov_prod_bridge.weight),2), '9G999G999D99') as
    revenue
from fact
join movie on fact.fk_m = movie.m_id
join mov_prod_bridge on movie.m_id = mov_prod_bridge.m_id
```

```
join producer on mov_prod_bridge.p_id = producer.p_id
where producer.studio <> 'Not␣Known'
group by producer.studio
order by revenue desc
limit 20;
```

Listing 4.2 SQL statement for selecting the revenue per producer studio

Table 4.2 Result set of Example 21

Studio	Revenue
Warner Bros. Pictures	3,395,622.45
Paramount Pictures	2,791,120.27
Universal Pictures	2,113,640.23
20th Century Fox	1,506,394.33
Walt Disney Pictures	1,411,725.65
DreamWorks SKG	1,301,846.83
Columbia Pictures Corporation	878,258.58
New Line Cinema	823,352.75
Columbia Pictures	815,260.67
Pixar Animation Studios	741,381.50
Marvel Enterprises	638,750.25
DreamWorks Animation	579,009.00
Metro-Goldwyn-Mayer	546,663.17
United Artists	507,799.58
Touchstone Pictures	476,309.42
Miramax Films	454,652.17
Apatow Productions	438,710.50
Imagine Entertainment	380,606.58
Walt Disney Company	368,420.50
Amblin Entertainment	365,394.50

The result is shown in Table 4.2.

Example 22. In order to offer personalized selections to customers, the movie rental company analyzes the user rating of a movie based on the age of the customers. Using this analysis, the company will then propose highly rated movies to customers based on the age group the customer belongs to. Customers are grouped into three classes: old, middle and young as follows:

- Old: Customers older than 40
- Middle: Customers between 20 and 40
- Young: Customers younger than 20

The proposition should contain the five best rated movies.

This analysis can be executed with a slice operation as illustrated below. The grouping of the customer into classes is a formatting of the result set and is not obvious in the cube C. In order to slice the five best-rated movies, the user rating has to be ranked according to customer class. For each customer class, the user rating is ordered from lowest to highest. From this ordered set of rankings, the highest five user rankings are extracted. In the standard SQL syntax [ISO08], the window function rank() allows partitioning and ranking of data sets based on an attribute. Therefore, the top 5 ranked movies can be extracted by selecting the movies with the higher rank than $max(rank(user\_rating)) - 4$. The slice is as follows

$slice\ (C = << movie.name,\ customer.bday >,\ user\_rating,\ R >,\ rank \geq max\ (rank(user\_rating)) - 4)$

The SQL statement for the slice operation is divided into four nested queries. The innermost nested query uses a case statement to group the age of the customers into old, middle and young. The next query applies the rank() function to the user ratings for each customer group. The third query adds the maximum ranking value for the five movies. In the outermost query, the top 5 movies for each customer class are selected. The SQL statement is as follows:

```
select name as "movie", cag as "customer␣age␣group"
from (
select b.name, b.cag , b.rank, (max(b.rank) over (partition by b
    .cag) -4) as top
From (
select a.name, a.cag, rank()  over (partition by a.cag order by
    a.rating) as rank
from (
select movie.name as name,
case
when extract(year from age(customer.bday)) >= 40
then 'old'
when extract(year from age(customer.bday)) >= 20 and extract(
    year from age(customer.bday)) < 40
then 'middle'
else 'young'
end as cag,
avg(fact.user_rating) as rating
from fact
join movie on fact.fk_mont = movie.pk_month
join customer on fact.fk_customer = customer.pk_customer
group by  movie.name, cag
) a
group by a.name, a.cag, a.rating
) b
group by b.name, b.cag, b.rank
) c
where c.rank >= c.top
```

Listing 4.3 SQL statement for selecting the customer age group per movie

Table 4.3 Result set of Example 22

Movie	Customer age group
Jackass 3	Middle
The Rugrats Movie	Middle
Witness for the Prosecution	Middle
The Lion King	Middle
Modern Times	Middle
The Great Dictator	Old
The Night of the Hunter	Old
Modern Times	Old
Witness for the Prosecution	Old
Network	Old
Toy Story 3	Young
The Lion King	Young
Jackass 3	Young
The Rugrats Movie	Young
The Incredibles	Young

The result set is as follows:

The analysis in Example 22 has several limitations. The classification of customer is crisp. A 40 year old customer might have similar taste in movies as a 39 year old customer, yet the 40 year old customer is rated differently. This behavior can be observed when limiting the SQL statement of Example 22 to the customer age groups 39, 40 and 41. The 40 years old customers rate the top 5 movie congruent with the age group middle, even if they are classified as old. Compared to the customer age groups 39 and 41, the age 40 rates movies closer to the 39 age group. On the other hand, the customer age 41 rates congruent to the customer age group old. This example illustrates that when classifying crisply, elements on the boundaries of the classes are often classified incorrect (Table 4.3). The SQL statement in Listing 4.4 shows the query adapted in order to get the user rating of the customer age groups 39, 40 and 41. Table 4.4 shows the result of the query.

```
select name, cag as "customer␣age␣group"
from (
select b.name, b.cag , b.rank , (max(b.rank) over (partition by b
    .cag) -4) as top
From (
select a.name, a.cag, rank()  over (partition by a.cag order by
    a.rating) as rank
from (
select movie.name as name,
case
when extract(year from age(customer.bday)) = 41
then '41'
when extract(year from age(customer.bday)) = 40
then '40'
when extract(year from age(customer.bday)) = 39
then '39'
```

```
end as cag,
avg(fact.user_rating) as rating
from fact
join movie on fact.fk_movie = movie.pk_movie
join customer on fact.fk_customer = customer.pk_customer
where extract(year from age(customer.bday)) in (39, 40, 41)
group by  movie.name, cag
) a
group by a.name, a.cag, a.rating
) b
group by b.name, b.cag, b.rank
) c
where c.rank >= c.top
```

Listing 4.4 SQL statement of customer age groups 39–41

Table 4.4 Result set of customer age group 39–41

Movie	Customer age group
Toy Story 3	39
The Lion King	39
Modern Times	39
Witness for the Prosecution	39
The Rugrats Movie	39
The Rugrats Movie	40
Modern Times	40
Witness for the Prosecution	40
Jackass 3	40
The Lion King	40
The Night of the Hunter	41
Witness for the Prosecution	41
The Great Dictator	41
Arsenic and Old Lace	41
Modern Times	41

The same classification problem can be observed with the top 5 movie rating. Looking at the user ratings for customer group old, the top 8 rated movies for this age group vary from 9.22 to 9.62. The result set is shown in Table 4.5. The difference between the number 5 rated movie (Network) and the number 6 rated movie (The Deer Hunter) is 0.02. Considering this minor difference in the rating, customers belong to age group old might be as interested in the movie Network as in the movie The Deer Hunter. Therefore, the limitation of five movies is only partly adequate to provide a movie list for suggesting to the customer.

The analysis and the corresponding query is rather complex as it nests four subqueries. It is probable that over time the classification will be adapted to new needs. In the customer age group young, the top 20 movies are presented and in the age group old, only the top 3 movies are presented. Therefore, adapting the query is a major effort. It might be simpler to extract the user ratings per age and do

Table 4.5 User rating of top 8 movies of customer age group old

Movie	Customer age group	User rating
Witness for the Prosecution	Old	9.6288767123287671
Modern Times	Old	9.4651467505241090
The Night of the Hunter	Old	9.4120500782472613
The Great Dictator	Old	9.3853645556146887
Network	Old	9.2529539295392954
The Deer Hunter	Old	9.2390846922672278
The Exorcist	Old	9.2338066630256690
The Rugrats Movie	Old	9.2256530825496343

the classification outside the data warehouse. At least when a classification should have multiple names for its classes, the cost of integrating all requirements in a single query are expensive and therefore the classification should be done outside. The consequence of outsourcing the classification is that only a restricted user group might have access to the classification. In order to improve such applications of the data warehouse, it can be enhanced to a fuzzy data warehouse.

4.2 Integration of Fuzzy Concepts in the Data Warehouse

In addition to the classical analysis in a data warehouse, the movie rental company needs some features that are available using fuzzy concepts.

4.2.1 Dimension Movie

The movies are classified into different genres. In the classical data warehouse, a movie always belongs fully to one or more genres, which is represented in the hierarchy path from movie to category in the movie dimension. In reality, movies can often be categorized into several genres while belonging more to one genre than to another. For instance, the movies Star Wars can be classified Science Fiction and Action, but more strongly Science Fiction. The movie Spaceballs, which is a parody of Star Wars, can be classified as Science Fiction and more strongly as Comedy.

With the classification in the classical data warehouse approach, the movie rental company cannot classify the movies asynchronously into different genres. Therefore, the company classifies the movies with a fuzzy concept. Based on the method presented in Sect. 3.3, first the category attribute movie is defined as target attribute. The second step is to identify the linguistic terms. In this case, the linguistic terms are the different genres to which the movies belong. These genres can be extracted from the dimension hierarchy category in the movie dimension. The company identifies one set of linguistic terms. Hence, this case corresponds to case 1 of the second task. In the third task, the membership functions for each genre has to be defined. The membership functions are discrete functions, respectively step functions [MWAS05], and a movie can belong with five different degrees to a genre $\mu_{genre} = [0, 0.33, 0.5, 0.66, 1]$. This definition of the membership functions corresponds to case 1 of task 3 in the modeling method in Sect. 3.3. The company should classify every movie manually as it is not possible to derive the degree of belonging to a genre from existing data. A manual classification of all the 661 movies demands initially a major effort, but considering that movies are instances of a category attribute of a dimension, this effort has to be done only once. Category attributes of dimensions in a data warehouse in general are static elements and are not often modified. For newly entered movies, the classification has still to be done once manually.

After identifying target attributes, linguistic terms and their membership functions, the meta table structure for the fuzzy concept can be defined based on the second step of the modeling method. According to the first and second task, one fuzzy classification table and one fuzzy membership table have to be created. The fuzzy classification table holds the genres as class membership attributes. The fuzzy membership table contains membership degrees for each target attribute corresponding to class membership attributes. The size of the fuzzy membership table is a Cartesian product of the movie table and fuzzy classification table. For the third and final task, the fuzzy classification table, the fuzzy membership table and the target attribute table have to be related to each other. This final task is completed by relating the foreign key attributes of the fuzzy membership table with the primary key attributes of target attribute table and of the fuzzy classification table.

By using the proposed approach, sharp and fuzzy classification of the movies in the data warehouse is maintained, which makes it possible to analyze the classification in both ways, sharp and fuzzy manners. On the other hand, it is possible to remove the category table and the corresponding bridge table as the fuzzy concept is redundant to this dimension hierarchy path and both classification have to be maintained independently to each other. Figure 4.2 shows the table schema of the dimension movie and the fuzzy concept movie genre.

The movie rental company further defines that the fuzzy concept movie genre can be aggregated with an arithmetic average function to dimension level producer. In comparison, the crisp classification defined as dimension hierarchy path does not allow to classify the movie producers. There, the movie is either rolled-up to producer or to movie category.

4.2.2 Dimension Customer

The movie rental company is interested in analyzing the revenue based on customer classifications. A common classification is an ABC analysis in which customers realizing 80 % of the revenue are classified as A category customers and all other customers are classified in B or C category. Most of the customers rent movies regularly and the achieved revenues per customer are close to each other. Therefore,

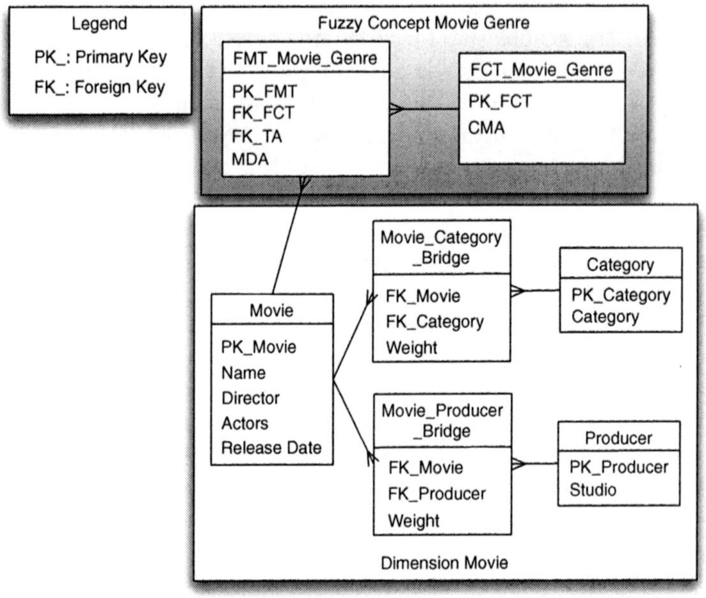

Fig. 4.2 Fuzzy concept movie genre

the classical ABC analysis does not provide a satisfying classification for the movie rental company. Using a fuzzy concept that classifies customers into good, moderate and bad customers is a more suitable classification. The fuzzy concept allows a more fine grained classification of customers. The customer category attribute in dimension customer is defined as target attribute in the first task of the modeling method. In the second task, one set of linguistic terms is defined as $\{good, moderate, bad\}$. An adaptive fuzzy concept is used and the membership functions are defined in the third task as follows:

$$lcr = lowest\ customer\ revenue$$

$$hcr = highest\ customer\ revenue$$

$$cr = customer\ revenue$$

$$base = hcr - lcr$$

$$\mu_{good}(cr) = \begin{cases} cr - lcr \geq 0.8 \times base, & MD_{cr} = 1 \\ cr - lcr \leq 0.6 \times base, & MD_{cr} = 0 \\ else, & MD_{cr} = \frac{cr - lcr - 0.6 \times base}{0.8 \times base - 0.6 \times base} \end{cases}$$

$$\mu_{moderate}(cr) = \begin{cases} cr - lcr \geq 0.8 \times base, & MD_{cr} = 0 \\ cr - lcr \leq 0.2 \times base, & MD_{cr} = 0 \\ 0.4 \times base \leq cr - lcr \leq 0.6 \times base, & MD_{cr} = 1 \\ 0.2 \times base \leq cr - lcr \leq 0.4 \times base, & MD_{cr} = \frac{cr - lcr - 0.2 \times base}{0.4 \times base - 0.2 \times base} \\ else, & MD_{cr} = \frac{0.8 \times base - cr - lcr}{0.8 \times base - 0.6 \times base} \end{cases}$$

$$\mu_{bad}(cr) = \begin{cases} cr - lcr \geq 0.4 \times base, & MD_{cr} = 0 \\ cr - lcr \leq 0.2 \times base, & MD_{cr} = 1 \\ else, & MD_{cr} = \frac{0.4 \times base - cr - lcr}{0.4 \times base - 0.6 \times base} \end{cases}$$

The base is calculated by subtracting the lowest revenue from the highest revenue. The revenue per customer is dependent on the fact revenue. Therefore, every time a new fact instance of revenue is added, the customer revenue is modified too. Recalculating the fuzzy concept customer revenue on every change of the fact table would impact the performance of the fuzzy data warehouse in a negative way. Because of this, the fuzzy concept customer revenue is adapted on a monthly basis by the movie rental company.

Additionally, the customers are classified according to their age in order to analyze revenue based on customer age. The target attribute is the category attribute customer as before. The customers are classified with a fuzzy concept containing the linguistic terms old, middle and young (second task of the modeling method). In the third task, the company defines the membership function as follows:

$$\mu_{old}(customer\ age) = \begin{cases} customer\ age \geq 65, & MD_{customer\ age} = 1 \\ customer\ age \leq 40, & MD_{customer\ age} = 0 \\ else, & MD_{customer\ age} = \frac{customer\ age - 40}{65 - 40} \end{cases}$$

$$\mu_{middle}(customer\ age) = \begin{cases} customer\ age \geq 65, & MD_{customer\ age} = 0 \\ customer\ age \leq 20, & MD_{customer\ age} = 0 \\ 30 \leq customer\ age \leq 40, & MD_{customer\ age} = 1 \\ 20 \leq customer\ age \leq 30, & MD_{customer\ age} = \frac{customer\ age - 20}{30 - 20} \\ else, & MD_{customer\ age} = \frac{65 - customer\ age}{65 - 40} \end{cases}$$

$$\mu_{young}(customer\ age) = \begin{cases} customer\ age \geq 30, & MD_{customer\ age} = 0 \\ customer\ age \leq 20, & MD_{customer\ age} = 1 \\ else, & MD_{customer\ age} = \frac{30 - customer\ age}{30 - 20} \end{cases}$$

This fuzzy concept can be characterized as an open end fuzzy concept. All customers older than 65 fully belong to the linguistic term old whereas customers younger 20 fully belong to the linguistic term young. Although the majority of customers are between 20 and 65 years old, it is still possible to adequately classify customers older than 65 or younger than 25.

In the first task of the second step of the modeling method, two fuzzy classification tables are created for the fuzzy concepts, and a fuzzy membership table for each fuzzy concept is created. Each fuzzy membership table will be the size

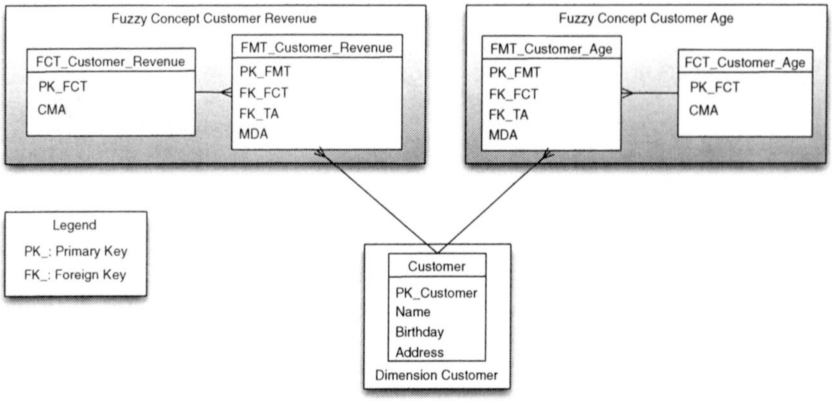

Fig. 4.3 Fuzzy concepts customer revenue and customer age

of the Cartesian product of the fuzzy classification table and the customer table. In the third step, the target attribute table customer, the fuzzy membership tables and the fuzzy classification tables are related to each other using the foreign key relations. Figure 4.3 illustrates the table schema of the dimension customer and the fuzzy concepts customer revenue and customer age.

4.2.3 Dimension Employee

The company is also interested analyzing the age of their employees. An employee age concept, similar to the fuzzy concept customer age, is created. The target attribute is the employee in the employee dimension (first task of first step in modeling method), the linguistic terms are old, middle and young (second task). Due to the nature of employee age, a limited fuzzy concept is used. The company has a policy that restricts employing people younger than 18. By Swiss federal law, retirement starts at the age of 65. In contrast to the fuzzy concept of customer age, the fuzzy concept employee age will never classify a person older than 65 or younger than 18. The membership functions are defined as follows (third task):

$$\mu_{old}(employee\ age) = \begin{cases} employee\ age \geq 45, & MD_{employee\ age} = 1 \\ employee\ age \leq 30, & MD_{employee\ age} = 0 \\ else, & MD_{employee\ age} = \frac{employee\ age - 30}{40-30} \end{cases}$$

$$\mu_{middle}(employee\ age) = \begin{cases} employee\ age \geq 45, & MD_{employee\ age} = 0 \\ employee\ age \leq 25, & MD_{employee\ age} = 0 \\ 30 \leq employee\ age \leq 35, & MD_{employee\ age} = 1 \\ 25 \leq employee\ age \leq 30, & MD_{employee\ age} = \frac{employee\ age - 25}{30-25} \\ else, & MD_{employee\ age} = \frac{45 - employee\ age}{45-30} \end{cases}$$

Fig. 4.4 Fuzzy concept employee age

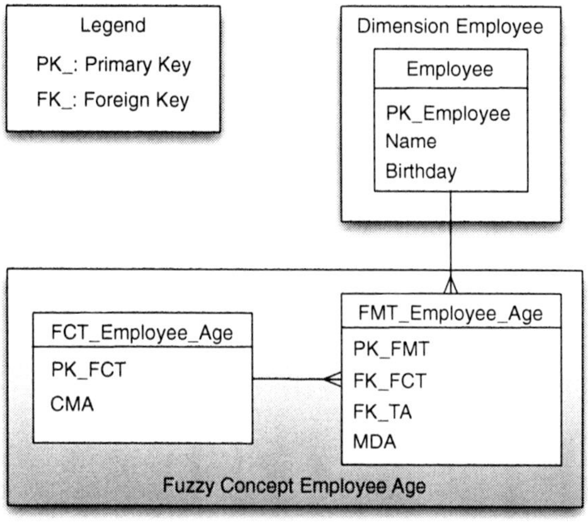

$$\mu_{young}(employee\ age) = \begin{cases} employee\ age \geq 30, & MD_{employee\ age} = 0 \\ employee\ age \leq 25, & MD_{employee\ age} = 1 \\ else, & MD_{employee\ age} = \frac{30 - employee\ age}{30 - 25} \end{cases}$$

To build the fuzzy concept employee age, one fuzzy classification table (first task of second step in modeling method) and one fuzzy membership table (second task) are created. In this case, the membership table might not be the size of the Cartesian product of the employee table and the fuzzy classification table. Considering that the movie rental company is hiring students during the summer break, it might be possible to have employees younger than 18. These employees are stored in the employee table but discarded by the fuzzy concept. Therefore, no entries are made in the fuzzy membership table for these employees. Finally, the customer table, the fuzzy membership table and the fuzzy classification table are related using the foreign key relation. Figure 4.4 presents the table schema of the fuzzy concept employee age.

4.2.4 Dimension Store

The surface for each store is registered in the data warehouse. The movie rental company is interested knowing the sales performance of big and small stores. Therefore, the company creates a fuzzy concept for classifying the store surface in big, medium and small. For the first task in the first step, the table store is chosen as target attribute and for the second task the linguistic terms big, medium and small are chosen. The fuzzy concept has the characteristic of a limited fuzzy concept as the company will not rent stores with a surface smaller than $40\,m^2$ or bigger than $250\,m^2$. For task 3, the membership functions are defined as follows:

$$\mu_{big}(store\ surface) = \begin{cases} store\ surface \geq 200, & MD_{store\ surface} = 1 \\ store\ surface \leq 150, & MD_{store\ surface} = 0 \\ else, & MD_{store\ surface} = \frac{store\ surface - 150}{200-150} \end{cases}$$

$$\mu_{medium}(store\ surface) = \begin{cases} store\ surface \geq 200, & MD_{store\ surface} = 0 \\ store\ surface \leq 80, & MD_{store\ surface} = 0 \\ 120 \leq store\ surface \leq 150, & MD_{store\ surface} = 1 \\ 80 \leq store\ surface \leq 120, & MD_{store\ surface} = \frac{store\ surface - 80}{120-80} \\ else, & MD_{store\ surface} = \frac{200 - store\ surface}{200-150} \end{cases}$$

$$\mu_{small}(store\ surface) = \begin{cases} store\ surface \geq 120, & MD_{store\ surface} = 0 \\ store\ surface \leq 80, & MD_{store\ surface} = 1 \\ else, & MD_{store\ surface} = \frac{120 - store\ surface}{120-80} \end{cases}$$

For the second step of the modeling method, the fuzzy classification table from task 1 and the fuzzy membership table from task 2 are defined to integrate this fuzzy concept in the fuzzy data warehouse. Differently from the employee age fuzzy concept, the fuzzy membership table of this limited concept has the size of the Cartesian product of the fuzzy classification and the store table. This is due to the fact that the company will always rent new stores in this range of surface, so all stores are classified. The relation of the tables is established by the foreign key relations. Figure 4.5 presents the table schema of the fuzzy concept store surface.

The fuzzy concept store surface can be propagated to the dimension levels city and region. The propagated fuzzy concepts are adaptive fuzzy concepts and the membership functions calculate the membership degrees based on the specific domain of the target attributes. Therefore, the membership function uses a ranking on the ordered set of the domain of attributes. The membership functions are as follows:

$$S = \{x_1, \ldots, x_n\} \ : \ \forall x_j \in dom_{TA} \wedge 1 \leq j \leq n \wedge x_j \leq x_{j+1}$$
$$rank(x_j) = \frac{j-1}{|S|-1}$$

$$\mu_{big}(store\ surface) = \begin{cases} rank(store\ surface) \geq 0.8, & MD_{store\ surface} = 1 \\ rank(store\ surface) \leq 0.6, & MD_{store\ surface} = 0 \\ else, & MD_{store\ surface} = \frac{rank(store\ surface) - 0.6}{0.8-0.6} \end{cases}$$

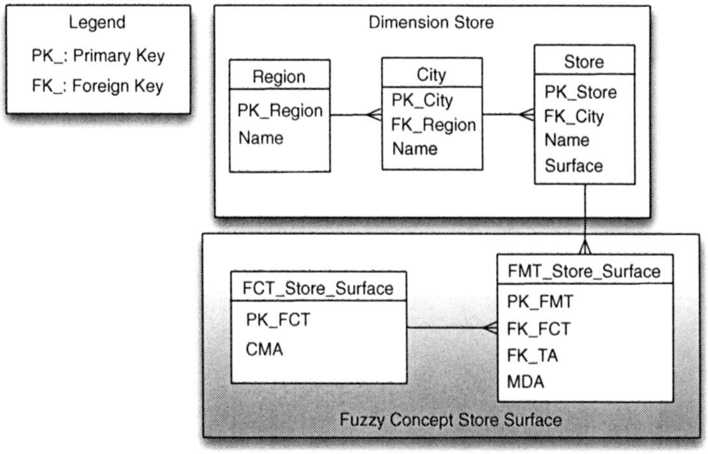

Fig. 4.5 Fuzzy concept store surface

$$\mu_{medium}(store\ surface) = \begin{cases} rank(store\ surface) \geq 0.8, & MD_{store\ surface} = 0 \\ rank(store\ surface) \leq 0.2, & MD_{store\ surface} = 0 \\ 0.4 \leq rank(store\ surface) \leq 0.6, & MD_{store\ surface} = 1 \\ 0.2 < rank(store\ surface) < 0.4, & MD_{store\ surface} = \frac{rank(store\ surface)-0.2}{0.4-0.2} \\ else, & MD_{store\ surface} = \frac{0.8-rank(store\ surface)}{0.8-0.6} \end{cases}$$

$$\mu_{small}(store\ surface) = \begin{cases} rank(store\ surface) \geq 0.4, & MD_{store\ surface} = 0 \\ srank(store\ surface) \leq 0.2, & MD_{store\ surface} = 1 \\ else, & MD_{store\ surface} = \frac{0.4-rank(store\ surface)}{0.4-0.3} \end{cases}$$

The propagated fuzzy concepts reuse the fuzzy classification table. Hence, for each dimension level on which the fuzzy concept is propagated, an additional fuzzy membership table has to be defined. Figure 4.6 illustrates the fuzzy concept store surface with the propagated fuzzy concepts on level city and region. The fuzzy membership tables for the propagated fuzzy concepts are represented as dashed table objects.

4.2.5 *Fact Revenue*

The movie rental company also creates a classification on the revenue in the fact table. According to the modeling method, the fact table is chosen in the first task of step 1 as the target attribute. Revenue should be classified as high, middle and

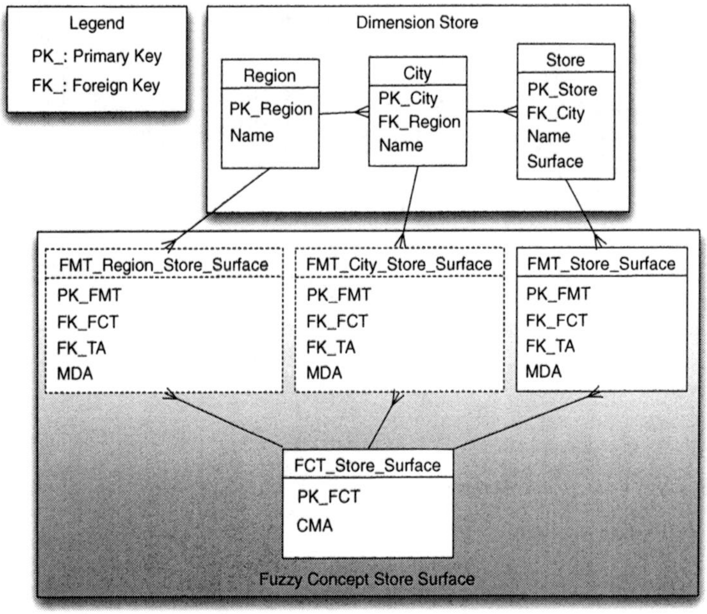

Fig. 4.6 Propagated fuzzy concepts city store surface and region store surface

low. In task 2, the linguistic terms high, middle and low are defined. Renting a movie costs either 5 or 7 CHF a day. The revenue is defined by the cost of renting a movie times the rental duration. Consequently, the revenue is a discrete value and always a multiple of 5 or 7. An open end fuzzy concept with discrete membership functions can be used in this case. In task 3, the membership functions are defined as follows:

$$\mu_{high}(revenue) = \begin{cases} revenue \geq 25, \ MD_{revenue} = 1 \\ revenue \leq 14, \ MD_{revenue} = 0 \\ revenue = 15, \ MD_{revenue} = 0.33 \\ revenue = 20, \ MD_{revenue} = 0.5 \\ revenue = 21, \ MD_{revenue} = 0.66 \end{cases}$$

$$\mu_{middle}(revenue) = \begin{cases} revenue = 15, & MD_{revenue} = 1 \\ revenue = 21, & MD_{revenue} = 0.5 \\ 25 \leq revenue \leq 10, & MD_{revenue} = 0 \\ revenue = 14, & MD_{revenue} = 0.5 \end{cases}$$

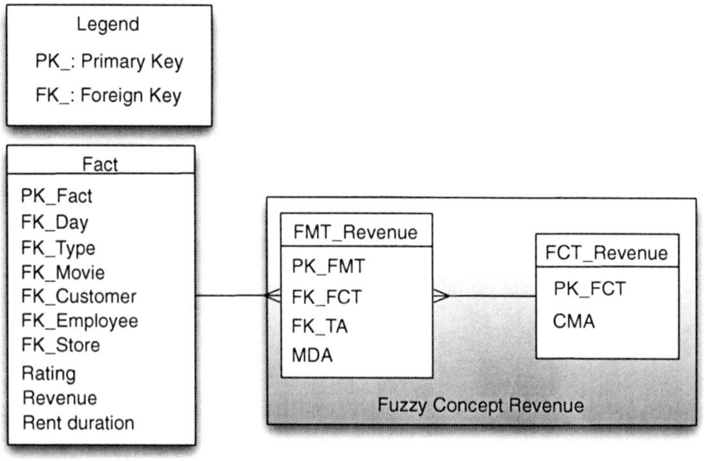

Fig. 4.7 Fuzzy concept revenue

$$\mu_{low}(revenue) = \begin{cases} revenue = 5, & MD_{revenue} = 1 \\ revenue = 7, & MD_{revenue} = 0.66 \\ revenue = 10, & MD_{revenue} = 0.33 \\ revenue \geq 14, & MD_{revenue} = 0 \end{cases}$$

In the second step of the modeling method, one fuzzy classification table for task 1 and one fuzzy membership table for task 2 are defined. The foreign keys of the fuzzy membership table are related with the primary keys of the fact table and the fuzzy classification table in order to build the relations in step 3. Figure 4.7 presents the table schema of the fuzzy concept revenue.

The fuzzy concept revenue is an important fuzzy concept for the movie rental company. Therefore, it is propagated to all the category attributes of every dimension. In total, 13 category attributes exist in the data warehouse and for each category attribute a fuzzy membership table has to be created. The propagated fuzzy concepts are defined as adaptive fuzzy concepts and the membership functions are, as in Sect. 4.2.4 described, a ranking of the domain of the target attributes. The membership functions for all propagated fuzzy concepts are as follows:

$$S = \{x_1, \ldots, x_n\} \; : \; \forall x_j \in dom_{TA} \wedge 1 \leq j \leq n \wedge x_j \leq x_{j+1}$$
$$rank(x_j) = \frac{j-1}{|S|-1}$$

$$\mu_{big}(revenue) = \begin{cases} rank(revenue) \geq 0.8, & MD_{revenue} = 1 \\ rank(revenue) \leq 0.6, & MD_{revenue} = 0 \\ else, & MD_{revenue} = \frac{rank(revenue)-0.6}{0.8-0.6} \end{cases}$$

$$\mu_{medium}(revenue) = \begin{cases} rank(revenue) \geq 0.8, & MD_{revenue} = 0 \\ rank(revenue) \leq 0.2, & MD_{revenue} = 0 \\ 0.4 \leq rank(revenue) \leq 0.6, & MD_{revenue} = 1 \\ 0.2 < rank(revenue) < 0.4, & MD_{revenue} = \frac{rank(revenue)-0.2}{0.4-0.2} \\ else, & MD_{revenue} = \frac{0.8-rank(revenue)}{0.8-0.6} \end{cases}$$

$$\mu_{small}(revenue) = \begin{cases} rank(revenue) \geq 0.4, & MD_{revenue} = 0 \\ srank(revenue) \leq 0.2, & MD_{revenue} = 1 \\ else, & MD_{revenue} = \frac{0.4-rank(revenue)}{0.4-0.3} \end{cases}$$

It must be noted that the customer revenue fuzzy concept, discussed in Sect. 4.2.2 for the customer dimension, is not the same fuzzy concept. The linguistic terms and the membership functions are not identical. Therefore, the movie rental company decides to leave the fuzzy concept customer revenue as a distinct fuzzy concept instead of redefining it as a propagated fuzzy concept. The fuzzy concept revenue is still propagated on the dimension customer. The dimension customer can hence be classified by two different revenue concepts.

The target attribute revenue of the base fuzzy concept revenue is modified as soon as new data is added into the data warehouse. The actual data from the source systems is loaded every night by the ETL process of the data warehouse. The classification of the propagated fuzzy concept revenue has to be recalculated as soon this event takes place in order to always guarantee a realtime classification. The movie rental company decides to trigger the recalculation of the fuzzy concept revenue after every daily load.

Figure 4.8 shows the base fuzzy concept revenue and the propagated fuzzy concepts. For simplicity the relations between the fact table and the dimensions are omitted.

Furthermore, the company requires that the fuzzy concept revenue can be propagated on every calculated cube that contains the measure revenue. Calculated cubes are the result of analyzing the fuzzy data warehouse and they only exist during the query process of the analysis. Therefore, the resulting revenue is volatile and the classification of the revenue has to be realized during the query process (see Sect. 3.4.3). For the instant calculation of the fuzzy concept, the stored procedure as described in Sect. 3.4.5 might be used.

4.2.6 Fact User Rating

Customers are asked to rate every movie when they return it. The rating of the movie is captured in the fact table as a decimal number between 0 and

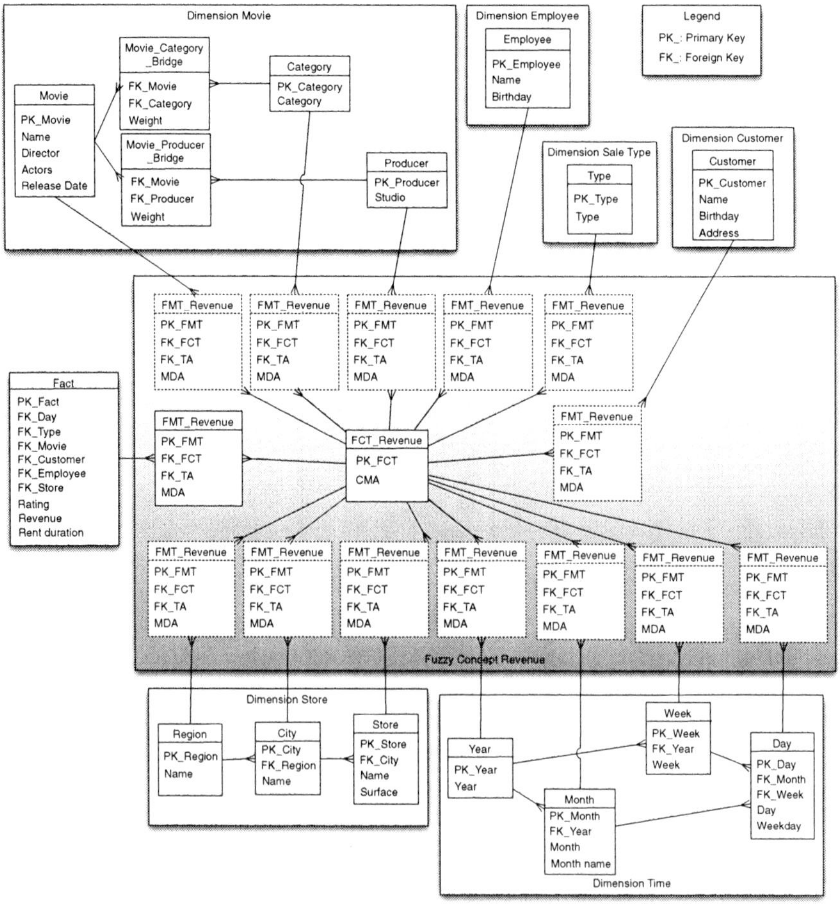

Fig. 4.8 Propagated fuzzy concepts revenue

10. If a customer does not rate the movie NULL is applied to the fact entry. The rating is used to create the company's particular list of hit movies. This list is presented to the customers on display screens in the movie stores as an additional service. The movie rental company uses this customer rating to classify the movies fuzzily into very good, good and ok movies. For this fuzzy concept, the target attribute is the user rating attribute in the fact table (task 1 in first step of the modeling method) and the linguistic terms are very good, good, and ok (task 2). As the rating is always between 0 and 10 a limited fuzzy concept can be created. The membership functions are the following (task 3):

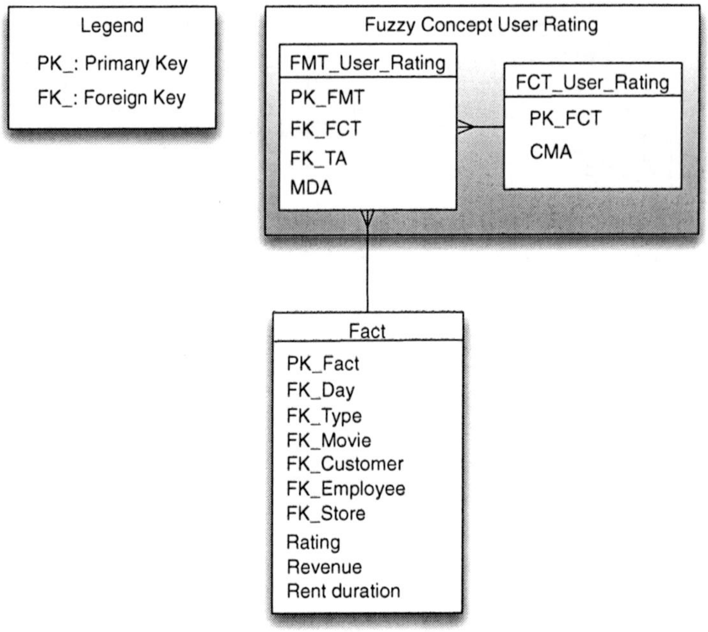

Fig. 4.9 Fuzzy concept user rating

$$\mu_{very\_good}(rating) = \begin{cases} rating \geq 8.0, & MD_{rating} = 1 \\ rating \leq 6.0, & MD_{rating} = 0 \\ else & MD_{rating} = \frac{rating-6.0}{8.0-6.0} \end{cases}$$

$$\mu_{good}(rating) = \begin{cases} rating \geq 8.0, & MD_{rating} = 0 \\ rating \leq 2.0, & MD_{rating} = 0 \\ 4.0 \leq rating \leq 6.0, & MD_{rating} = 1 \\ 2.0 \leq rating \leq 4.0, & MD_{rating} = \frac{rating-2.0}{4.0-2.0} \\ else & MD_{rating} = \frac{8.0-rating}{8.0-6.0} \end{cases}$$

$$\mu_{ok}(rating) = \begin{cases} rating \geq 4.0, & MD_{rating} = 0 \\ rating \leq 2.0, & MD_{rating} = 1 \\ else & MD_{rating} = \frac{4.0-rating}{4.0-2.0} \end{cases}$$

To integrate this concept in the fuzzy data warehouse, a fuzzy classification table (task 1 of the second step) and a fuzzy membership table (task 2) are defined and related to the fact table (task 3). Figure 4.9 presents the table schema of the fuzzy concept user rating.

The fuzzy concept rating might only be aggregated over the movie and customer dimensions as the fact user rating only provides meaningful information in conjunction with these two dimensions. The aggregation function of the crisp fact user rating is an average function: the user rating on level movie is the arithmetic average of

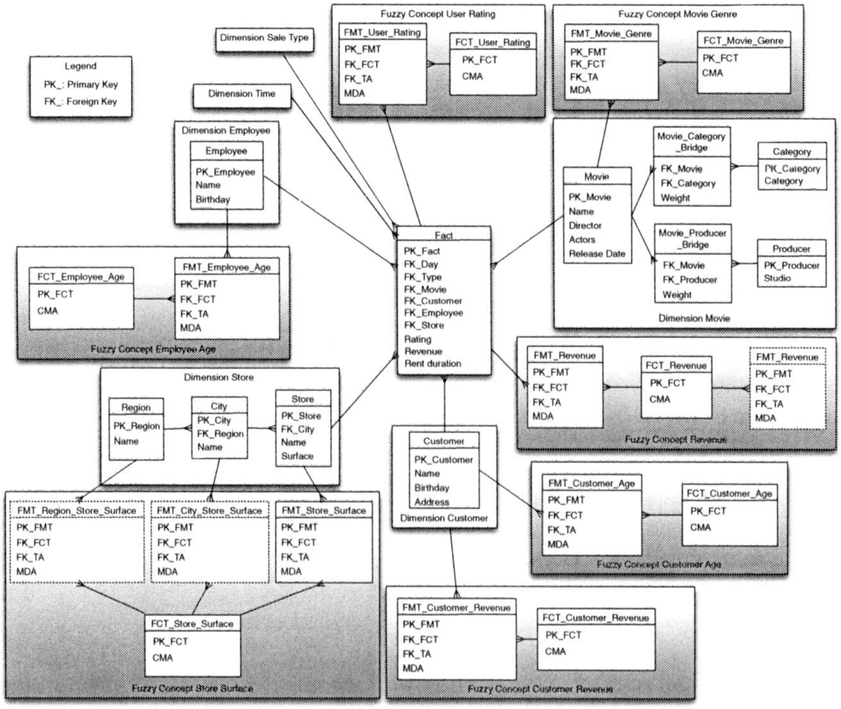

Fig. 4.10 Fuzzy data warehouse schema for the movie rental company

all the registered ratings in the fact table for a specific movie. Similarly, the ratings for the levels producer and category on the dimension movie and for level client on dimension client can be deducted. Subsequently, the user ratings for each level are aggregated from the lower dimension level.

4.2.7 *Fuzzy Data Warehouse Schema*

Finally, the data warehouse schema is extended with the meta tables of the fuzzy concepts. In Fig. 4.10 the fuzzy data warehouse schema is presented including the meta tables for all fuzzy concepts shaded grey. For each fuzzy concept, the fuzzy classification table and the fuzzy membership tables are presented. The linguistic terms for a fuzzy concept are stored in the attribute CMA in the fuzzy classification table and the membership degrees of a target attribute to the corresponding linguistic term is stored in the attribute MDA in the fuzzy membership table. The fuzzy membership tables of propagated fuzzy concepts are presented as dashed fuzzy membership table objects within the fuzzy concepts.

For readability, the 13 propagated fuzzy membership tables of the fuzzy concept revenue are only symbolized by 1 fuzzy membership table object and the relations to the category attributes are omitted. For simplicity, the dimensions which do not have any fuzzy concept are only depicted as boxes with the dimension name. The structures of these simplified dimensions are retained as shown in Fig. 4.1.

4.3 Using the Fuzzy Data Warehouse

After creating the fuzzy data warehouse, the movie rental company can now take advantage of the fuzzy concepts for analysis. In this section, how the movie rental company can use the fuzzy concepts in order to improve their analysis is discussed.

Example 23. The movie rental company is interested in analyzing how much revenue was generated by older customers. To determine this, a fuzzy slice operation on the fuzzy data warehouse can be executed. In this operation, the fuzzy concept customer age is used as slicer. The result set of the slice operation can be further refined with an α-cut. The slice operation is executed as follows:

slice(<< customer.customer.name, customer.customer.age >, < customer.customer.customer_age.old >, revenue, R>, customer.customer.customer_age.old > 0.8)

The SQL statement of this slice operation is shown in Listing 4.5.

```
select customer.name as Name, extract("year" from age(bday)) as
    Age, fmt.mda as "CMA␣old", sum(revenue) as Revenue
from fact
join ( customer
join ( fmt_customer_age as fmt
join fct_customer_age as fct on fmt.fk_fct = fct.pk_fct and fct.
    cma = 'old') on fmt.fk_ta = customer.pk_customer) on fact.
    fk_customer = customer.pk_customer
where fmt.mda > 0.8
group by Name, Age, "CMA␣old"
order by Revenue
```

Listing 4.5 SQL statement for selecting the revenue of old customers

Table 4.6 shows the result set of the ten first customers.

Example 24. The company is interested in the distribution of revenues of movie rentals by the age of the customers. To answer this question, the company applies the fuzzy concept customer age as selector. In order to get a result that is consistent with the overall revenue, the membership degrees of old, middle and young for each target attribute instance has to be normalized to 1. Otherwise, the sum of the

Table 4.6 Result set of Example 23

Name	Age	CMA old	Revenue
Kai Fischer	70	1.00	2,335
Alissa Pfeiffer	63	0.92	2,338
Jannis Vondembussche-haddenhausen	65	1.00	2,397
Maria Fink	65	1.00	2,424
Leon Hinkel	76	1.00	2,432
Pepe Petri	74	1.00	2,443
Danny Thomae	68	1.00	2,480
Carina Lumm	71	1.00	2,493
Clemens Schlagdenhaufen	69	1.00	2,513
Jean Bensing	63	0.88	2,523

Table 4.7 Result set of Example 24

CMA	Revenue
Old	17,183,705.52
Middle	20,944,178.78
Young	10,055,692.70

revenue of all linguistic terms will not correspond to the actual overall revenue. The cube operation and the fuzzy selector are as follows:

$<<>, < customer.customer.customer\_age >, revenue \times customer.customer.customer\_age.mda, R >$

The corresponding SQL statement is in Listing 4.6.

```
select fct.cma as "CMA", sum(revenue * fmt.mda) as Revenue
from fact
join ( customer
join ( fmt_custage as fmt
join fct_custage as fct on fmt.fk_fct = fct.pk_fct )
on customer.pk_customer = fmt.fk_ta )
on customer.pk_customer = fact.fk_customer
group by "CMA"
```

Listing 4.6 SQL statement for selecting the revenue per customer age group

The result set is presented in Table 4.7.

Example 25. The company also wants to analyze how much revenue small stores are producing. The time frame July to August 2010 is chosen to reduce the size of the result set. In addition, the data is grouped by region. For this analysis, the following dice operation can be used:

dice($<< time.month, region.region, region.store.name, region.store.surface >, < region.store.\_store\_surface.small >, revenue, R >, \{time.year.year{=}2010, time.$

month.month = "*July*", *time.month.month* = "*August*", *region.store.store_surface* = "*small*"}, {*AND, AND, AND*})

Only the revenue of the stores with a small store surface is aggregated to the dimension hierarchy level region. Hence, small store surface and not region store surface is a slicer of the dice operation. The other slicers are the time restrictions July, August and 2010. The corresponding SQL statement is shown in Listing 4.7.

```
select "month".monthname as "Month", region.region as "Region",
    store.name as "Store", store.surface as "Surface", fmt.mda
    as "CMA_small", sum(revenue) as "Revenue"
from fact
join ( "day"
join ( "month"
join "year" on "month".fk_year = "year".pk_year and "year"."year
    " = '2010' )
on "day".fk_month = "month".pk_month )
on fact.fk_day = "day".pk_day
join ( store
join ( fmt_storesurf as fmt
join fct_storesurf as fct on fmt.fk_fct = fct.pk_fct and fct.cma
    = 'small' and fmt.mda > 0.00) on fmt.fk_ta = store.pk_store
join ( city
join region on region.pk_region = city.fk_region )
on store.fk_city = city.pk_city )
on store.pk_store = fact.fk_store
group by "Month", "Region", "Store", "Surface", "CMA_small"
order by "Revenue"
```

Listing 4.7 SQL statement for selecting revenue of small stores

Contrary to the cube operation in Example 24, the fuzzy concept in this dice operation is not involved in the summation process of revenue. It is only involved as slicer in order to list the stores with small store surface. Hence, in this case the membership degrees do not necessarily have to be normalized.

A shortened result set of the first ten data rows is presented in Table 4.8.

In Example 22, the movie rental company tried to classify movies based on the user rating of certain age groups. This classification was intended to be used as a personalized suggestion to customers. The user in a certain age group should then receive a list of the top 5 rated movies of this age group. However, the crisp classification of age and the crisp classification of movies are not adequate to provide this personalized offer. Furthermore, the classification in Example 22 implies a rather complex nested SQL statement that is not flexible for further extension of the classification. Therefore, the classification with a third party tool was proposed. The next example will provide a similar classification based on the fuzzy concepts in order to show how fuzzy concepts can facilitate such classification.

Table 4.8 Result set of Example 25

Month	Region	Store	Surface	CMA small	Revenue
February	French	Store LAC2	100	0.50	16,528
February	German	Store LUZ1	90	0.75	16,543
February	French	Store GEN2	90	0.75	17,978
February	German	Store THU1	55	1.00	18,065
February	French	Store LAU1	90	0.75	18,328
June	German	Store LUZ1	90	0.75	18,347
February	German	Store THU2	65	1.00	18,751
November	German	Store BER1	75	1.00	18,759
April	German	Store LUZ1	90	0.75	18,765
August	French	Store SIO1	105	0.38	18,999

Example 26. The aim of the classification is to provide customers with a personalized list of movies they might be interested in. Therefore, the company classifies the movies according the user rating of a certain age group. Only the highest rated movies should then be shown in the list. This analysis can be completed by classifying the customers with the fuzzy concept customer age and the movie with the aggregated fuzzy concept user rating. Instead of proposing the top 5 movies, the analysis can use an $\alpha - cut$ on fuzzy concept user rating to only propose the best classified movies. This analysis can be executed using a fuzzy slice operation in which the slicer is the fuzzy concepts user rating on movie. The slice is as follows:

slice(<< movie.movie,customer.customer >, < customer.customer.customer_age, movie.movie.user_rating.very_good >,revenue,R >, movie.movie.user_rating.very_good > 0.94)

The corresponding SQL statement for this slice is given in Listing 4.8.

```
select movie.name as "Movie", a.custcma as "Customer␣age␣group",
       round(avg(a.mda),2) as "Rating"
from fact
join ( movie
join (select fctc.cma as custcma, pk_movie as ta, avg(
     fmt_user_rating.mda) as mda
from fact
join ( customer
join ( fmt_customer_age as fmtc
join fct_customer_age as fctc on fctc.pk_fct = fmtc.fk_fct and
     fmtc.mda > 0.00)
on customer.pk_customer = fmtc.fk_ta )
on customer.pk_customer = fact.fk_customer
join fmt_user_rating on fact.pk_fact = fmt_user_rating.fk_ta
join fct_user_rating on fmt_user_rating.fk_fct = fct_rating.
     pf_fct and fct_rating.cma = 'very␣good'
join movie on fact.fk_movie = movie.pk_movie
group by custcma, ta
```

```
having avg(fmt_user_rating.mda) >= 0.94) a
on movie.pk_movie = a.ta )
on fact.fk_movie = movie.pk_movie
group by "Customer␣age␣group", "Movie"
order by "Customer␣age␣group", "Movie"
```

Listing 4.8 SQL statement for selecting movie rating

In contrast to the SQL statement in Example 22, only one nested query is used. The nested query groups the fuzzy concept user rating based on the fuzzy concept customer age and aggregates it to the category attribute movie. This nested query is only necessary when aggregation of the fuzzy concepts is not executed in a prior step by the fuzzy data warehouse. Nevertheless, the query in the fuzzy data warehouse is less complex than the corresponding query in the classical data warehouse. One major reason for the simplification of the query is the fact that the logic of the classification of age group and top 5 movies is not directly integrated in the query. The fuzzy concepts user rating and customer age implement this logic and the fuzzy data warehouse only has to combine the corresponding meta tables in order integrate the classification. The result set of this query contains 27 data rows, in comparison to the 15 data rows of result set in Example 22. Table 4.9 presents a shortened result set of 15 data rows.

Using the fuzzy concept user rating does not limit the analysis to five movies per age group. With the $\alpha - cut$, the movie rental company has the possibility to define how strong the rating should belong to the desired class, here into very good rating. In contrast, the classification in Example 22 chooses the five best rated movies without regard to the actual rating of the movies. When the sixth movie is rated as high as the fifth one, it is still not considered. The query order might be influenced by the name of the movie or the internal query optimizer of the data warehouse system – the crisp selection of the movies is not only taking the user rating into consideration. Whereas, the $\alpha - cut$ on the fuzzy concept user rating chooses every movie belonging to the class very good movie by more than 0.94 without being restricted by the query order. This effect is the reason the number of movies vairies by the different age groups, i.e. the age group old has rated more movies as very good than the age group middle.

The main advantage of using the fuzzy concept user rating instead of the crisp user rating is the pre-grouping of the ratings before applying the classification. The company can use a linguistic term in order to choose only the very good rated movies. Using the crisp ratings would end in a restriction based on the numeric values of the user rating, i.e every rating higher 8.0, which would correspond to an $\alpha - cut$ of 1.0 of the linguistic term very good. The fuzzy concept allows a global definition of what a very good movie is according the rating. Therefore, when using fuzzy concept in multiple analysis, the meaning of the classification remains equal.

When analyzing the age groups, it can be noted that the user rating per age group considers every rating of a customer that belongs some degree more than 0.0 to that class. A customer with age 45 influences the customer age groups old and middle

Table 4.9 Result set of Example 26

Movie	Customer age group	Rating
Modern Times	Middle	0.96
The Great Dictator	Middle	0.95
The Night of the Hunter	Middle	0.95
The Rugrats Movie	Middle	0.96
Witness for the Prosecution	Middle	0.98
Harvey	Old	0.94
Ikiru	Old	0.94
...	Old	...
The Rugrats Movie	Old	0.96
Witness for the Prosecution	Old	0.99
Infernal Affairs	Young	0.95
Jackass 3	Young	0.97
The Incredibles	Young	0.95
...	Young	...
The Lion King	Young	0.97

as 45 belongs 0.8 to the middle age and 0.2 to the old age group. The movie rating classification does not take the different membership degrees of age into account. The rating of the 45 year old customer is fully considered for the two age groups. The user rating and its fuzzy concept are aggregated using the arithmetic average, and movie rating is influenced by a much broader number of customers than the crisp rating in Example 22. It consequently allows smoother transitions between the age groups and the customers at the borderline are classified more accurately with respect to their real rating behavior.

The movie rental company has the flexibility to restrict the number of customer ages that are taken into account by increasing the $\alpha - cut$ of the customer age classification. The $\alpha - cut$ in the current SQL statement of Example 26 is 0.0. When increasing it to 1.0, only ratings of customers that fully belong to an age group are considered. The company has the option to use the $\alpha - cuts$ to fine tune their classification in a way that a crisp classification can not provide it.

With the fuzzy classification in Example 26, customer can receive a personalized list of movie propositions according their belonging to the age groups. Considering a 45 years old customer, who belongs to the age groups middle age (0.8) and old age (0.2), the list can be composed of movies in both age groups. According to the level of belonging to the groups, a specific subset of movies can be proposed in order to further personalize the service. For example, the company might propose to the 45 year old customers the 20 % best rated movies in age group old and the 80 % best rated movies of the age group middle. This further personalization is only possible when customers can belong to more than one age group at a time and is therefore not possible to achieve with crisp classification of customers.

Example 27. Next, the movie rental company wants to analyze rentals of movies and rentals over time. The time period will be restricted to every month of year 2010. The company is not only interested in the amount of revenue a movie produced during a particular time period, but also how strongly the movie performs in overall revenue. Additionally, the overall performance of the period should be analyzed too. In order to combine all these needs into a single report, the company is combining the two fuzzy concepts movie revenue and monthly revenue. The analysis can be completed by slicing the fuzzy cube $<$ *movie.movie, time.year, time.month* $>$, $<$ *movie.movie.revenue, time.month.revenue* $>$, *revenue*, $R >$ with the slicer 2010. The resulting operation is as follows:

slice($<$ *movie.movie, time.year,time.month* $>$, $<$ *movie.movie.revenue, time.month.revenue* $>$, *revenue, R* $>$, *time.year* = 2010)

The corresponding SQL statement is listed in 4.9.

```
select "month".monthname, morevh.mda as "Month␣CMA␣high", morevm
    .mda as "Month␣CMA␣middle", morevl.mda as "Month␣CMA␣low",
    movie.name, mrevh.mda as "Movie␣CMA␣high", mrevm.mda as "
    Movie␣CMA␣middle", mrevl.mda as "Movie␣CMA␣low", sum(revenue
    ) as "Revenue"
from fact
join ("day"
join ("month"
join (fmt_revenue_month as "morevh"
join fct_revenue as "fctmorevh" on morevh.fk_fct = fctmorevh.
    pk_fct and fctmorevh.cma = 'high') on "month".pk_month =
    morevh.fk_ta
join (fmt_revenue_month as "morevm"
join fct_revenue as "fctmorevm" on morevm.fk_fct = fctmorevm.
    pk_fct and fctmorevm.cma = 'middle') on "month".pk_month =
    morevm.fk_ta
join (fmt_revenue_month as "morevl"
join fct_revenue as "fctmorevl" on morevl.fk_fct = fctmorevl.
    pk_fct and fctmorevl.cma = 'low') on "month".pk_month =
    morevl.fk_ta
join "year" on month.fk_year = "year".pk_year) on "day".fk_month
    = "month".pk_month ) on fact.fk_day = "day".pk_day
join ( movie
join (fmt_movrev as "mrevh"
join fct_revenue as "fctmrevh" on fctmrevh.pk_fct = mrevh.fk_fct
    and fctmrevh.cma = 'high') on movie.pk_movie = mrevh.fk_ta
join (fmt_movrev as "mrevm"
join fct_revenue as "fctmrevm" on fctmrevm.pk_fct = mrevm.fk_fct
    and fctmrevm.cma = 'middle') on movie.pk_movie = mrevm.
    fk_ta
join (fmt_movrev as "mrevl"
join fct_revenue as "fctmrevl" on fctmrevl.pk_fct = mrevl.fk_fct
    and fctmrevl.cma = 'low') on movie.pk_movie = mrevl.fk_ta )
    on fact.fk_movie = movie.pk_movie
where "year"."year" = 2010
group by "month".monthname, morevh.mda, morevm.mda, morevl.mda,
    movie.name, mrevh.mda , mrevm.mda, mrevl.mda
order by sum(revenue) desc
```

Listing 4.9 SQL statement for selecting movie revenue

In Table 4.10, the results with the top 10 revenues of the report are presented. It should be noted that the highest revenues are not necessarily produced in the best rated periods or with the highest rated movies; only the month July is belonging with 1.0 to high monthly revenue.

The same picture is presented when analyzing the ten lowest revenues earned by movies in a period in 2010, which is presented in Table 4.11.

In contrast to an analysis of a sharp cube presenting the movie revenues over the months in 2010, the fuzzy cube directly provides information about the popularity of movies and periods based on their overall revenue. In order to get similar information by using sharp cubes, at least three cubes have to be queried and combined first. Next to the cube resulting from the slice, two cubes resulting from the roll-ups of the revenue over the movies and revenue over month have to be executed. Further, the composed sharp cube has to be classified twice: once according to movie revenue and once according to monthly revenues. The classifications might be done in a fuzzy manner, resulting in the same cube as the single fuzzy cube operation provided above. Still, the classifications are only executed during the querying process of the analysis and wont be persistently available. Considering the fact that the classification might be used several times in different analysis, the fuzzy data warehouse provides a persistent classification which is valid for all analysis and reduces the calculation overhead as the classification is only calculated once.

Example 28. The movie rental company can reuse the fuzzy cube from Example 27 and classify the revenue of the cube. The fuzzy concept, defined on the fact revenue, can be propagated on non persistent cubes as defined in Sect. 4.2.5. In order to complete this classification, the adaptive fuzzy concept revenue is calculated during the execution of the query. The operation for adding the fuzzy concept to the cube is as follows:

$$fuzzify(slice(< movie.movie,\ time.year,\ time.month >, < movie.movie.revenue, time.month.revenue >, revenue, R >, time.year = 2010), revenue)$$

Considering the calculation of the volatile fuzzy concept is completed with a stored procedure as explained in Sect. 3.4.5, the SQL statements for the fuzzyfying operation are as follows:

```
select temp.fc_revenue('select␣month.monthname,␣morevh.mda␣as␣"
    Month␣CMA␣high",␣morevm.mda␣as␣"Month␣CMA␣middle",␣morevl.
    mda␣as␣"Month␣CMA␣low",␣movie.name,␣mrevh.mda␣as␣"Movie␣CMA␣
    high",␣mrevm.mda␣as␣"Movie␣CMA␣middle",␣mrevl.mda␣as␣"Movie␣
    CMA␣low",␣sum(revenue)␣as␣"Revenue"
from␣fact
join␣(day
join␣(month
join␣(fmt_revenue_month␣as␣"morevh"
join␣fct_revenue␣as␣"fctmorevh"␣on␣morevh.fk_fct␣=␣fctmorevh.
    pk_fct␣and␣fctmorevh.cma␣=␣'high')␣on␣month.pk_month␣=␣
    morevh.fk_ta
join␣(fmt_revenue_month␣as␣"morevm"
```

```
join␣fct_revenue␣as␣"fctmorevm"␣on␣morevm.fk_fct␣=␣fctmorevm.
    pk_fct␣and␣fctmorevm.cma␣=␣'middle')␣on␣month.pk_month␣=␣
    morevm.fk_ta
join␣(fmt_revenue_month␣as␣"morevl"
join␣fct_revenue␣as␣"fctmorevl"␣on␣morevl.fk_fct␣=␣fctmorevl.
    pk_fct␣and␣fctmorevl.cma␣=␣'low')␣on␣month.pk_month␣=␣morevl
    .fk_ta
join␣year␣on␣month.fk_year␣=␣year.pk_year)␣on␣day.fk_month␣=␣
    month.pk_month␣)␣on␣fact.fk_day␣=␣day.pk_day
join␣(␣movie
join␣(fmt_movrev␣as␣"mrevh"
join␣fct_revenue␣as␣"fctmrevh"␣on␣fctmrevh.pk_fct␣=␣mrevh.fk_fct
    ␣and␣fctmrevh.cma␣=␣'high')␣on␣movie.pk_movie␣=␣mrevh.fk_ta
join␣(fmt_movrev␣as␣"mrevm"
join␣fct_revenue␣as␣"fctmrevm"␣on␣fctmrevm.pk_fct␣=␣mrevm.fk_fct
    ␣and␣fctmrevm.cma␣=␣'middle')␣on␣movie.pk_movie␣=␣mrevm.
    fk_ta
join␣(fmt_movrev␣as␣"mrevl"
join␣fct_revenue␣as␣"fctmrevl"␣on␣fctmrevl.pk_fct␣=␣mrevl.fk_fct
    ␣and␣fctmrevl.cma␣=␣'low')␣on␣movie.pk_movie␣=␣mrevl.fk_ta␣)
    ␣on␣fact.fk_movie␣=␣movie.pk_movie
where␣year.year␣=␣2010
group␣by␣month.monthname,␣morevh.mda,␣morevm.mda,␣morevl.mda,␣
    movie.name,␣mrevh.mda,␣mrevm.mda,␣mrevl.mda
order␣by␣sum(revenue)␣desc');
select * from temp.revvol;
```

Listing 4.10 SQL statement for fuzzyfying a volatile cube

Table 4.12 shows the result set of the first ten entries ordered by month.

With a single fuzzy concept, the movie rental company classified three different revenues. Each classification provides a view of another value range of revenue while still using the same fuzzy concept definition. In Example 28, the movie rental company not only sees how good a movie or a month performed on overall revenue, but also sees the performance of a movie during a specific month. In a further step, the company can extend the analysis by starting to combine the fuzzy concepts. For example, the movie company has to remove a certain number of movies from its stock for 2011. Deciding which movie has to be removed based only on the overall performance of the movie might force the removal of movies that performed well in 2010. This might be seen as indicator that the movie will also perform well in 2011. Taking the classification of month into account, the company can analyze if some movies performed better on weak months. Such movies might be rented more frequently during periods that have weaker overall performance (ex: Christmas movies during winter time or summer night movies during summer time). Therefore, these movies might improve revenue during a certain period but compared with the overall performance they still might be classified as moderate or weak. Such movies can be more easily identified when combining the different revenue fuzzy concepts and might improve the decision-making process for which movie should be removed in 2011.

Table 4.10 Ten highest revenues by movies per month in 2010

Month name	Month CMA high	Month CMA middle	Month CMA low	Movie name	Movie CMA high	Movie CMA middle	Movie CMA low	Revenue
January	0.70	0.30	0.00	Transformers: Revenge of the Fallen	0.82	0.18	0.00	10,479
January	0.70	0.30	0.00	Cloudy with a Chance of Meatballs	0.57	0.43	0.00	10,213
January	0.70	0.30	0.00	The Twilight Saga: New Moon	0.32	0.68	0.00	10,136
January	0.70	0.30	0.00	It's Complicated	0.29	0.71	0.00	10,066
January	0.70	0.30	0.00	Harry Potter and the Half-Blood Prince	0.70	0.30	0.00	9,989
March	0.00	1.00	0.00	The Secret in Their Eyes	0.62	0.38	0.00	9,982
January	0.70	0.30	0.00	The Blind Side	0.32	0.68	0.00	9,891
January	0.70	0.30	0.00	Fast & Furious	1.00	0.00	0.00	9,877
March	0.00	1.00	0.00	Avatar	0.26	0.74	0.00	9,842
January	0.70	0.30	0.00	The Proposal	0.89	0.11	0.00	9,758

Table 4.11 Ten lowest revenues by movies per month in 2010

Month name	Month CMA high	Month CMA middle	Month CMA low	Movie name	Movie CMA high	Movie CMA middle	Movie CMA low	Revenue
October	0.36	0.64	0.00	Double Jeopardy	0.00	0.49	0.51	235
February	0.00	0.00	1.00	Toy Story 2	0.00	0.41	0.59	235
August	0.56	0.44	0.00	Batman Forever	0.00	0.42	0.58	235
June	0.00	0.00	1.00	Hotaru no haka	0.00	0.39	0.61	230
October	0.36	0.64	0.00	Lilo & Stitch	0.00	0.49	0.51	225
October	0.36	0.64	0.00	The World Is Not Enough	0.00	0.43	0.57	225
November	0.00	1.00	0.00	Yôjinbô	0.00	0.46	0.54	225
October	0.36	0.64	0.00	Barry Lyndon	0.00	0.44	0.56	220
December	1.00	0.00	0.00	The Pianist	0.00	0.41	0.59	215
August	0.56	0.44	0.00	There's Something About Mary	0.00	0.36	0.64	215

Table 4.12 First ten result sets of fuzzy cube of Example 28

Month name	Mo. high	Mo. middle	Mo. low	Movie name	Movie high	Movie middle	Movie low	Revenue	Revenue high	Revenue middle	Revenue low
January	0.70	0.30	0.00	101 Dalmatians	0.00	0.43	0.57	575	0.0	0.28	0.72
January	0.70	0.30	0.00	12 Angry Men	0.00	0.43	0.57	530	0.0	0.21	0.79
January	0.70	0.30	0.00	2 Fast 2 Furious	0.00	0.48	0.52	495	0.0	0.15	0.85
January	0.70	0.30	0.00	2001 A Space Odyssey	0.00	0.33	0.67	795	0.0	0.64	0.36
January	0.70	0.30	0.00	2012	0.32	0.68	0.00	8,988	1.0	0.0	0.0
January	0.70	0.30	0.00	300	0.60	0.40	0.00	545	0.0	0.23	0.77
January	0.70	0.30	0.00	50 First Dates	0.91	0.09	0.00	555	0.0	0.25	0.75
January	0.70	0.30	0.00	8 Mile	0.00	0.39	0.61	560	0.0	0.26	0.74
January	0.70	0.30	0.00	A Beautiful Mind	0.00	0.42	0.58	800	0.0	0.65	0.35
January	0.70	0.30	0.00	A Bug's Life	0.00	0.41	0.59	520	0.0	0.19	0.81

Concepts can also be combined to form new classifications of movies. Using the concept of hierarchical decomposition [Wer08], the three fuzzy concepts of the cube in Example 28 can be combined in order to form a new classification for movies. The company can group movies that performed similarly with regard to the overall performance, the performance in a specific month in 2010 and the monthly overall performance. This classification leads to a new portfolio classifying movies from top-top-top movies: movies that have high membership degrees in all fuzzy concepts; to worst-worst-worst movies. This portfolio not only provides better classification of movies because of smoother transitions between the classes, it also provides a single, global view of different revenue aspects per movie. This is due to the fuzzy concept revenue, a single classification of revenue, that can be applied to different value ranges (aspects) of revenue.

With the application of fuzzy concepts, the movie rental company improved its possibilities to analyze data in their data warehouse. Classifications are defined once and can then be consistently applied to different aggregation levels. Therefore, the classification and the subsequent interpretation of specific cubes are more traceable compared to classifications only based on a single result set of an analysis. Using the meta table structure of the fuzzy data warehouse, the company has the option to add variations of fuzzy concepts in terms of different linguistic terms or different membership functions. This gives the possibility to define classifications that takes different interpretations of business users into consideration. Finally, the fuzzy nature of the classifications allows smother transitions between classes and therefore allows more accurate classification than similar sharp approaches.

Part III
Implementation

Chapter 5
Implementation of Fuzzy Data Warehouse

Chapter 4 discussed the application of a fuzzy data warehouse for a movie rental company, and demonstrated how a fuzzy concept can be integrated in data analysis. The corresponding SQL statements and result set were shown. However, for end users, the application only provides direct access to the database system of the data warehouse. Therefore, the user has to know the structure of the meta tables and the data warehouse. This meta information can not be stored directly in the database system. In order to facilitate interaction with the fuzzy data warehouse, a proof of concept of a fuzzy data warehouse implementation is presented in this chapter. In addition to the raw database application, the implementation provides a business logic that contains the meta information about the table structures. Furthermore, a front end is created that allows easier interaction with the fuzzy data warehouse than typing pure SQL statement in a console.

The next sections describe the implementation of the proof of concept. In Sect. 5.1, an overview of the architecture is given. In this overview, the structure of the proof of concept is depicted and the dependencies between the different architectural layers are examined. The database tier is depicted in Sect. 5.2. Section 5.3 provides further information about the processing of the meta information and the engines that provide the communication between the first and third tiers. Finally, Sect. 5.4 discusses the visual layer of the proof of concepts and shows how the end user can interact with the fuzzy data warehouse.

5.1 Architectural Overview

The architecture of the prototype is based on a three-tier architecture [Eck95]. The fuzzy data warehouse described in Chap. 4 represents the database layer. It is built with PostgreSQL [Pos11a] on three schemas that separate the data warehouse tables, the meta tables of the fuzzy concept and tables that are temporary created on runtime.

D. Fasel, *Fuzzy Data Warehousing for Performance Measurement*, Fuzzy Management Methods, DOI 10.1007/978-3-319-04226-8_5,

The business logic acts as middle tier [Eck95] and fulfills three tasks. First, it uses the meta information to build the fuzzy data warehouse based on the tables of the database layer. This meta information provides information on how dimensions are constructed from tables, what kind of facts lie in the fact table, how the facts are aggregated and how the fuzzy concept is built from the meta table structure. Therefore, the meta information can be seen as the transposing logic from the database tables to the fuzzy data warehouse schema presented in Fig. 4.10. The second task is to interpret the user action from the visualization layer and transform it into SQL statements for querying the fuzzy data warehouse. The last task is to manage fuzzy concept that can be created, altered or deleted by users within the visualization layer.

The visualization layer provides the user interface to facilitate interaction with the fuzzy data warehouse. The user is able to query the fuzzy data warehouse in a simpler way than typing SQL statements and the corresponding result sets are presented in the visualization layer. Additionally, the user is able to create, modify and delete fuzzy concepts in the visualization layer.

Figure 5.1 presents an overview of the architecture of the implementation. The database layer contains the PostgreSQL database with its schemas. The PostgreSQL schemas, its tables and stored procedures are further depicted in Sect. 5.2. The business logic layer includes the meta information represented in Fig. 5.1 as "Fuzzy Concept Meta Information" and "Data Warehouse Information". Additionally, it comprises an engine for processing the queries and an engine for processing the fuzzy concept administration. In Sect. 5.3, the different components of the business logic layer are discussed in greater detail. The visual layer contains the navigation components and the fuzzy concept administration components. All components are discussed in Sect. 5.4. The arrows in Fig. 5.1 show the communication paths between the different components.

5.2 Database

The database consists of three schemas: *fdwh* , *meta* and *temp*. The *fdwh* schema holds the tables of the classical data warehouse. To be compliant with the snowflake schema, the tables are in third normal form. Consequently, for each category attribute of every dimension, a unique table is created. The *fdwh* schema includes 16 tables that are described in the snowflake schema in Fig. 4.1.

The schema *meta* is used to save the fuzzy concept meta tables. For each fuzzy concept described in Sect. 4.2, at least one fuzzy classification table and one fuzzy membership table are created therein. For propagated fuzzy concepts, such as the fuzzy concept revenue, an additional fuzzy membership table is created for each propagated level. In the case of the fuzzy concept revenue, a fuzzy membership table per category attribute of the dimensions is created. In total, it comprises 14 fuzzy membership tables and 1 fuzzy classification table as illustrated in Fig. 4.8. The schema *meta* includes 30 fuzzy concept meta tables in total, comprising all the

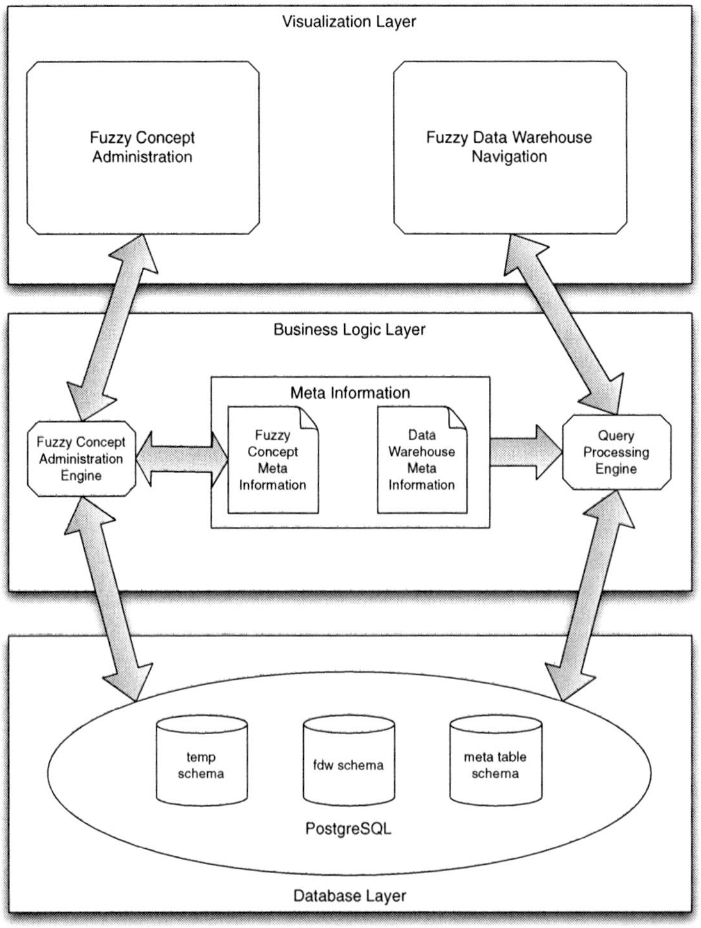

Fig. 5.1 Overview of prototype architecture

propagated fuzzy concepts. An overview of the tables, excluding the propagated tables of fuzzy concept revenue, is shown in the grey boxes of Fig. 4.10.

The temporary schema *temp* is used for tables that are built during classification of volatile target attributes. An example of how this schema can be used is given in the trigger illustrated in Listing 3.3. This trigger calculates the membership degrees of the fuzzy concept revenue for volatile target attributes. The SQL query of a cube having revenue as fact is passed to the trigger. The trigger then creates a temporary table in the schema temp that contains the attributes of the original cube plus an attribute for each class membership attribute of the fuzzy concept revenue. The SQL statement of the original cube is executed and its result set is saved in the table. In a further step, the membership degrees are calculated and stored in the temporary table. Finally, the temporary table is queried and the result set is returned to the user interface. It is worth noting that data in the tables of the *temp* schema can only be considered as correct at the time of the query. Temporary tables are not aware of changes in the original tables or in the *meta* schemas and therefore should always be purged after usage.

In addition to the tables and the schema, trigger functions and stored procedures are implemented in the database layer. For every fuzzy concept, at least one stored procedure is created in order to represent the membership function. In the role of the membership function, the stored procedure calculates the membership degree of every target attribute and class membership attribute. The membership degree is then stored directly in the fuzzy membership table. The trigger function monitors the target attribute table and calls the stored procedure each time the target attribute table changes. Based on the type of fuzzy concept used, the stored procedure calculates only the membership degree for the newly added or altered target attributes or, in case of an adaptive fuzzy concept, recalculates the membership degree of every target attribute. Trigger functions and stored procedures for the calculation of the fuzzy concept are always kept in the schema meta. Listing 5.1 below shows the stored procedure for the fuzzy concept store surface (see Sect. 4.2.4). The trigger meta.fc_storesurface_trigger shown in Listing 5.2 fires the stored procedure meta.fc_storesurf() after every row that is inserted or updated in the table store (see line 1–3 in Listing 5.2). As described in Sect. 4.2.4, the fuzzy concept store surface is a limited fuzzy concept and only holds membership degrees for surface values bigger than 40 and smaller than 250 m^2. The stored procedure implements the membership function as if-else clauses considering values between 40 and 250 (see line 11–35 in Listing 5.1). The if-else clauses insert the calculated membership degrees directly in the table fmt_storesurf, which is the fuzzy membership table of the fuzzy concept (see lines 12–14, 16–18, 20–22, 24–28 and 30–34 in Listing 5.1).

In addition to the trigger meta.fc_storesurface_trigger, two more triggers are created as shown in Listing 5.2 (lines 4–9). These triggers fire the stored procedures for the propagated fuzzy concepts city store surface and region store surface. The target attributes city and region store surface are aggregated values from store surface. Consequently, the triggers have to be created on table store. As opposed to the trigger for the fuzzy concept store surface, the stored procedure for the fuzzy concepts city and store surface are only triggered once after the insert statement. The reason for this can be seen in the stored procedure meta.fc_prop_storesurf() in Listing 5.3. The stored procedure aggregates all target attributes from the table store before calculating the corresponding membership degrees. Therefore, the stored procedure only has to be executed once and not for every inserted or updated row. As stated in Sect. 4.2.4, the fuzzy concepts city and region store surface are adaptive fuzzy concepts that calculate membership degree based on a ranking function. The aggregation and ranking of the surfaces are executed in a single nested SQL statement in the stored procedure (see line 14–19 in Listing 5.3). The calculated membership degrees are inserted into the tables meta.fmt_storesurf_city, respectively meta.fmt_storesurf_region, that represent the fuzzy membership tables of the propagated fuzzy concepts (see lines 23–24, 26–27, 29–30, 32–35 and 38–42 in Listing 5.3).

```
create or replace function meta.fc_storesurf() returns trigger as $$
declare
big integer; -- id of linguistic term big
medium integer; -- id of linguistic term medium
small integer; -- id of linguistic term small
membership numeric;
begin
select PK_fct into big from meta.fct_storesurf where cma = 'big';
select PK_fct into medium from meta.fct_storesurf where cma = 'medium';
select PK_fct into small from meta.fct_storesurf where cma = 'small';
if 250 < NEW.surface and NEW.surface >= 200 then
  insert into meta.fmt_storesurf (fk_fct, fk_ta, mda) values (big, NEW.PK_Store, 1.00);
  insert into meta.fmt_storesurf (fk_fct, fk_ta, mda) values (medium, NEW.PK_Store, 0.00);
  insert into meta.fmt_storesurf (fk_fct, fk_ta, mda) values (small, NEW.PK_Store, 0.00);
elsif NEW.surface <= 80 then
  insert into meta.fmt_storesurf (fk_fct, fk_ta, mda) values (big, NEW.PK_Store, 0.00);
  insert into meta.fmt_storesurf (fk_fct, fk_ta, mda) values (medium, NEW.PK_Store, 0.00);
  insert into meta.fmt_storesurf (fk_fct, fk_ta, mda) values (small, NEW.PK_Store, 1.00);
elsif 120 <= NEW.surface and NEW.surface <= 150 then
  insert into meta.fmt_storesurf (fk_fct, fk_ta, mda) values (big, NEW.PK_Store, 0.00);
  insert into meta.fmt_storesurf (fk_fct, fk_ta, mda) values (medium, NEW.PK_Store, 1.00);
  insert into meta.fmt_storesurf (fk_fct, fk_ta, mda) values (small, NEW.PK_Store, 0.00);
elsif 150 < NEW.surface and NEW.surface < 200 then
  membership := (NEW.surface - 150.00)/50.00;
  insert into meta.fmt_storesurf (fk_fct, fk_ta, mda) values (big, NEW.PK_Store, cast(membership as decimal(3,2)));
  membership := (200.00 - NEW.surface)/50.00;
  insert into meta.fmt_storesurf (fk_fct, fk_ta, mda) values (medium, NEW.PK_Store, cast(membership as decimal(3,2)));
  insert into meta.fmt_storesurf (fk_fct, fk_ta, mda) values (small, NEW.PK_Store, 0.00);
elsif 40 < NEW.surface then
  membership := (NEW.surface - 80.00)/40.00;
  insert into meta.fmt_storesurf (fk_fct, fk_ta, mda) values (big, NEW.PK_Store, 0.00);
  insert into meta.fmt_storesurf (fk_fct, fk_ta, mda) values (medium, NEW.PK_Store, cast(membership as decimal(3,2)));
  membership := (120.00- NEW.surface)/40.00;
  insert into meta.fmt_storesurf (fk_fct, fk_ta, mda) values (small, NEW.PK_Store, cast(membership as decimal(3,2)));
```

```
end if;
end;
$$ language plpgsql;
```

Listing 5.1 Stored procedure for fuzzy concept store surface

```
create or replace trigger meta.fc_storesurface_trigger after
    insert or update
on fdwh.store on each row
execute meta.fc_storesurf();
create or replace trigger meta.fc_city_storesurface_trigger
    after insert or update
on fdwh.store on each statement
execute meta.fc_prop_storesurf('city');
create or replace trigger meta.fc_region_storesurface_trigger
    after insert or update
on fdwh.store on each statement
execute meta.fc_prop_storesurf('region');
```

Listing 5.2 Triggers for fuzzy concept store surface

```
create or replace function meta.fc_prop_storesurf(text) returns
    trigger as $$
declare
level alias for $1;
query text;
big integer; -- id of linguistic term big
medium integer; -- id of linguistic term medium
small integer; -- id of linguistic term small
membership numeric;
surface record;
begin
select PK_fct into big from meta.fct_storesurf where cma = 'big'
    ;
select PK_fct into medium from meta.fct_storesurf where cma = '
    medium';
select PK_fct into small from meta.fct_storesurf where cma = '
    small';
if level = 'city' then
  query := 'select␣a.pk,␣percent_rank()␣over␣(order␣by␣a.surface
      )␣as␣rank␣from␣(select␣PK_' || level || '␣as␣pk,␣sum(store
      .surface)␣as␣surface␣from␣fdwh.store␣join␣fdwh.city␣on␣
      store.fk_city␣=␣city.PK_city)␣a';
elsif level = 'region' then
  query := 'select␣a.pk,␣percent_rank()␣over␣(order␣by␣a.surface
      )␣as␣rank␣from␣(select␣PK_' || level || ',␣sum(store.
      surface)␣as␣surface␣from␣fdwh.store␣join␣fdwh.city␣on␣
      store.fk_city␣=␣city.PK_city␣join␣fdwh.region␣on␣city.
      FK_region␣=␣region.PK_region)␣a';
else
  raise error;
end if;
```

```
for surface in execute query loop
if surface.rank >= 0.8 then
  query := 'insert␣into␣meta.fmt_storesurf_' || level ||'␣(
        fk_fct ,␣fk_ta ,␣mda)␣values␣(big ,␣' || surface.pk || ',␣
        1.00) ,␣(medium ,␣␣||␣surface.pk␣||␣', 0.00), (small , ||
        surface.pk || ',␣0.00)';
  execute query;
elsif surface.rank <= 0.2 then
 query := 'insert␣into␣meta.fmt_storesurf_' || level ||'␣(fk_fct
      ,␣fk_ta ,␣mda)␣values␣(big ,␣' || surface.pk || ',␣0.00) ,␣(
      medium ,␣␣||␣surface.pk␣||␣', 0.00), (small , || surface.pk
      || ',␣1.00)';
  execute query;
elsif 0.4 <= surface.rank and surface.rank <= 0.6 then
  query := 'insert␣into␣meta.fmt_storesurf_' || level ||'␣(
        fk_fct ,␣fk_ta ,␣mda)␣values␣(big ,␣' || surface.pk || ',␣
        0.00) ,␣(medium ,␣␣||␣surface.pk␣||␣', 1.00), (small , ||
        surface.pk || ',␣0.00)';
  execute query;
elsif 0.6 < surface.rank and surface.rank < 0.8 then
  membership := (surface.rank - 0.6)/0.2;
  query := 'insert␣into␣meta.fmt_storesurf_' || level ||'␣(
        fk_fct ,␣fk_ta ,␣mda)␣values␣(big ,␣' || surface.pk || ',␣
        cast(' || membership || '␣as␣decimal(3,2))) ,␣';
  membership := (0.8 - surface.rank)/0.2;
  query := query || '␣(medium ,␣' || surface.pk || ',␣cast(' ||
        membership || '␣as␣decimal(3,2))) ,␣(small ,␣surface.pk ,␣
        0.00)';
  execute query
else
  membership := (surface.rank - 0.2)/0.2;
  iquery := 'insert␣into␣meta.fmt_storesurf_' || level ||'␣(
        fk_fct ,␣fk_ta ,␣mda)␣values␣(big ,␣' || surface.pk || ',␣
        0.00) ,␣(medium ,␣' || surface.pk || ',␣cast(' || membership
          || '␣as␣decimal(3,2))) ,␣';
  membership := (0.4 - surface.rank)/0.2;
  query := query || '␣(small ,␣' || surface.pk || ',␣cast(' ||
        membership || '␣as␣decimal(3,2))) ,␣(small ,␣surface.pk ,␣
        0.00)';
  execute query;
end if;
end loop;
end;
$$ language plpgsql;
```

Listing 5.3 Stored procedure for propagated fuzzy concepts city and region store surface

While the logic for the membership functions could also be realized in the business logic layer as it is not directly a part of the pure data storage tasks, there are several advantages to execute this business logic on the database layer. First, the database engine has the most reliable information when a membership degree should be calculated for a target attribute. It is always aware of the state of its transaction and therefore it knows best when to trigger a stored procedure after a transaction. When calculating the membership degree on the database level, no additional communication with the business layer is necessary. Especially in environments where the database layer and the business logic layer are separated on different machines, communication between the tiers might be an performance issue. The most important factor in why stored procedures are chosen for implementing membership functions on the database layer is the fact that PostgreSQL provides a lot of functionalities for server programming [Pos11c]. Therefore, implementing membership functions using stored procedures takes advantage of the transactional system of the database engine. It can still be programmed with external tool such as Python or C-language code snippets where necessary.

5.3 Business Logic

The first task of the business logic is to provide meta information about the fuzzy data warehouse. This meta information contains the necessary information for transposing the raw database table structures into the fuzzy data warehouse schema. It has to describe how dimensions and fuzzy concepts are mapped in the table structure, how facts can be aggregated, and how fuzzy concepts are related to each other in case of propagated fuzzy concepts. The chosen approach for representing meta information is inspired by the open source data warehouse platform Mondrian [Pen10]. Mondrian uses a XSD schema document [HM05] in order to provide meta information. The main advantage of this approach is the ability to provide a framework of XML tags to precisely specify elements of the data warehouse. The tags can be further used to describe an instance of a data warehouse system. XML schemas allow defining well-formed schemas of a data warehouse in documents that are both user readable and interpretable by machines. Mondrian has not been chosen for the prototype implementation because it is a complete OLAP server platform and has many additional features that are not used in the proof of concept. These features, such as user right concepts, would increase the complexity of the prototype and would not be helpful for the objectives of the prototype: to show how the fuzzy data warehouse concept of Chap. 3 can be implemented.

5.3.1 XML Schema for Dimensions and Facts

The meta information of the prototype consists of a XML document describing the crisp elements of the fuzzy data warehouse and a XML document describing the fuzzy concepts. The fuzzy concepts are maintained in a separate XML document because the fuzzy concept administration engine will later modify this document. Therefore, with separation of the meta information into two documents, only data that must be modified is exposed to the fuzzy concept administration engine.

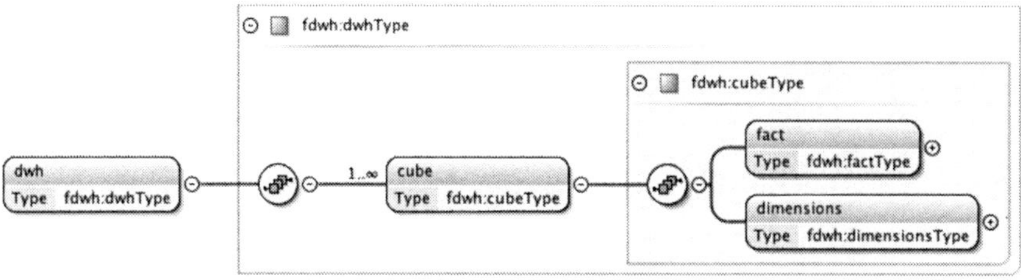

Fig. 5.2 Node relation "dwh", "cube", "fact" and "dimensions"

Separate XML schema documents are created for both XML documents. These documents describe the valid XML nodes in the documents and its structure. In the next paragraphs, the XML schema definitions are presented. The application of the schemas is illustrated by excerpts from the XML documents of the movie rental company fuzzy data warehouse. The complete XML schemas for the fuzzy data warehouse part can be found in Appendix A as well as in Appendix B. The XML documents describing the fuzzy data warehouse instance can be found in Appendices C and D.

The XML schema for describing the crisp elements of the fuzzy data warehouse includes a root node *dwh*. The root node can be composed of one or more *cube* nodes. A *cube* node represents one OLAP cube of the data warehouse. It might be possible to have a data warehouse with multiple OLAP cubes. In the example of the application in Chap. 4, only one OLAP cube is used to represent the fuzzy data warehouse. A *cube* node consists of a *fact* node and a *dimensions* node. Figure 5.2 illustrates the relation between the nodes *dwh*, *cube*, *fact* and *dimensions*.

The node *fact* holds the information about the fact table of the fuzzy data warehouse and the measures (facts) that are stored inside the fact table. Therefore, the node *fact* is composed of a node *relation* and one or more nodes *measure*. The node *relation* consists of a node *table*, a node *key* and a node *column*. The node *column* is optional and not used for describing the fact relation. The *table* node contains the name of the fact table and the *key* node holds the primary key attribute of the fact table. With this information, the *relation* node specifies the fact table on the database layer. For each measure of the fuzzy data warehouse, a *measure* node is created. A *measure* node contains the necessary information to identify a fact and its aggregation behavior. Hence, a sub-node *column* and a sub-node *aggregator* are included in the measure *node*. The *column* node specifies the attribute that holds the fact instances in the fact table. The *column* node is further divided into the nodes *column*, holding the name of the column, *table* node, and the optional node *display*. *Display* makes it possible to give another name to the fact than the column name. The node *aggregator* specifies how the fact is aggregated over the dimensions. The aggregation can be SUM, MAX, MIN, AVG, PROD. Figure 5.3 illustrates the *fact* node and its sub-node.

Listing 5.4 shows the instance of the *fact* node for the fuzzy data warehouse. The relation identifies the fact table "fact" with the primary key fact_id (lines 2–5

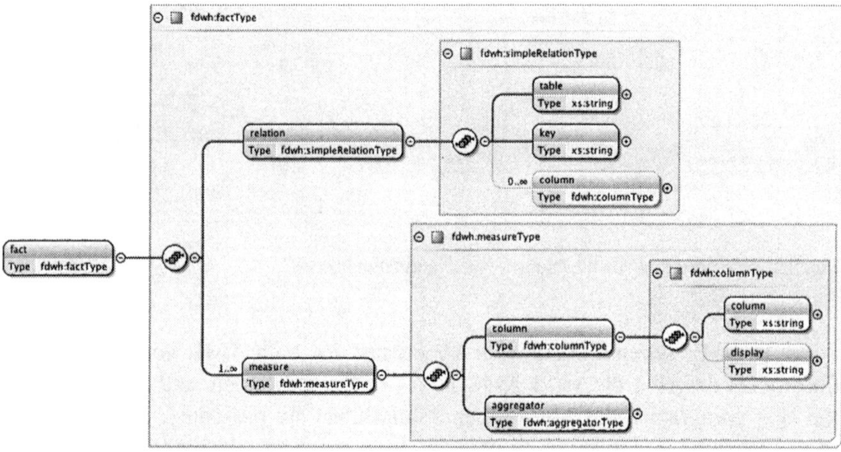

Fig. 5.3 Node "fact" and its children

in Listing 5.4). In the *measure* nodes, the two facts revenue (lines 6–11) and user rating (lines 12–18) are defined. It is notable that the fact user rating is defined in the fact table as user_rating but will be displayed as "User Rating" in the fuzzy data warehouse. Additionally, revenue is aggregated using a summation function and user rating using an average function.

```
<fact>
   <relation>
    <table>fact</table>
    <key>fact_id</key>
   </relation>
   <measure>
    <column>
     <column>revenue</column>
    </column>
    <aggregator>SUM</aggregator>
   </measure>
   <measure>
    <column>
     <column>user_rating</column>
     <display>User Rating</display>
    </column>
    <aggregator>AVG</aggregator>
   </measure>
</fact>
```

Listing 5.4 "Fact" node of fuzzy data warehouse

In order to be compliant with the data warehouse definitions of Kimball [KR02] and Inmon [Inm05] (see also Sect. 2.1), at least one time dimension and one additional dimension has to be specified in a data warehouse. Therefore, the node *dimensions* is composed of at least two dimensions each represented by a

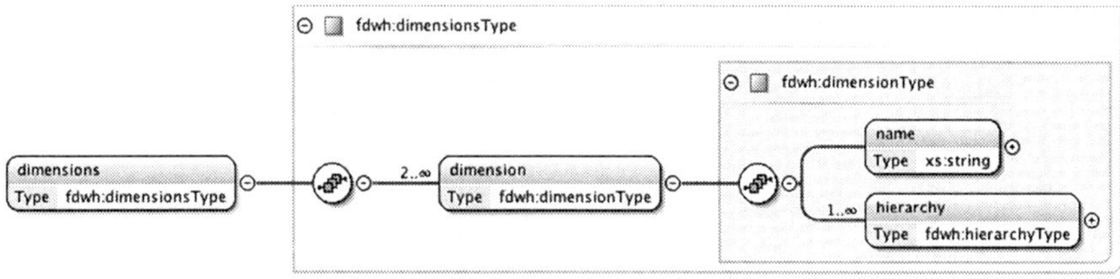

Fig. 5.4 Node "dimensions" and its children

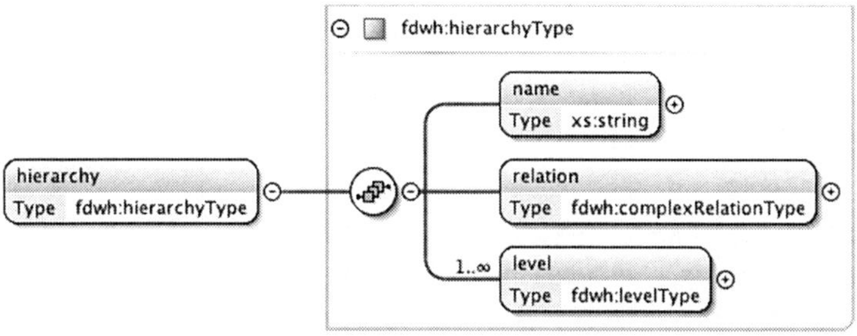

Fig. 5.5 Node "hierarchy" and its children

dimension node. The *dimension* node itself is composed of a node *name*, which provides the name of the dimension, and a node *hierarchy*. Figure 5.4 illustrates the nodes *dimensions*, *dimension* and the relation to its child nodes.

A dimension has at least one hierarchy over which a fact can be aggregated. The *hierarchy* node represents the hierarchies of the dimension. It is composed of the node *name*, *relation* and *level* as shown in Fig. 5.5. The node *name* holds the name of the hierarchy. The structure, how the hierarchy is stored in the database, is described within the node *relation*. The *relation* node is composed of a key *node* and either a *table*, *join* or *relation* node. Figure 5.6 illustrates the *relation* node and its children. The *relation* node can include other *relation* nodes, and can therefore construct the hierarchy structure iteratively. The uppermost parent *relation* node always holds another *relation* node and a *key* node that identifies the foreign key attribute in the fact table. Each subsequent *relation* node can be composed with one of the following variants:

- A *key* node and a *table* node. This variant can be used to specify a one-table relation. If the parent node of the *relation* is a *join* node and the *relation* is the last child of the parent node, then the *key* node holds the primary key of the table specified in the *table* node. Otherwise, the *key* node holds the foreign key that relates the table to the *key* node of the neighbor node.
- A *key* node and a *join* node. This variant can be used when the hierarchy is spread over multiple tables in the database. The *join* node is composed of exactly two *relation* nodes. The *relation* nodes themselves can be composed of one of the variants stated here. The *key* node specifies the primary key of the table

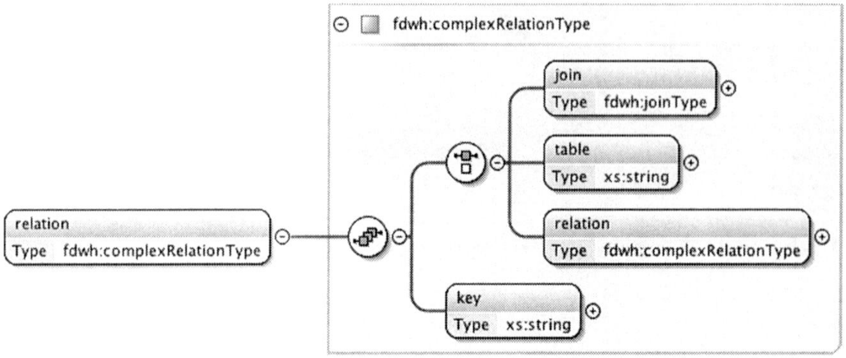

Fig. 5.6 Node "relation" and its children

in the first child node. Translated into an SQL statement, the first child would be the leftmost join table. Additionally, this *key* node provides the relation to the neighbor element of the parent.

- A *key* node and a *relation* node. A *relation* node always remains for another iteration step. The *key* node holds the primary key if the parent element is a *relation* node (except if the parent is the first relation in the node hierarchy) or if the parent element is a *join* node and the *relation* node is the last child. Otherwise, it holds a foreign key attribute.

In order to visualize these variants, the relation in hierarchy path time month of dimension time (see dimension time in Fig. 4.1) is presented in Listing 5.5. The category attribute at the highest level of dimension time is year. The *relation* node defined on lines 14–17 defines the corresponding table and primary key for year (type: key and table node). This relation is joined (capsuled in the *join* node from line 9 to 18) with the relation for the category attribute month (lines 10–13; type *key* and *join* node). The *key* node in relation month (line 12) is the foreign key attribute in table month that references the primary key attribute in table year (*key* node on line 16). The *join* node is further capsuled in a *relation* node (lines 8–20). According to the second variant above, the *key* node (line 19) in this *relation* node represents the primary key attribute of the table month. Next, the relation is joined within a *join* node (lines 3–21) and with another *relation* node (lines 4–7). The *relation* node on lines 4–7 contains the table specification for the category attribute day. As explained in the first variant, the *key* node on line 6 represents the foreign key in table day that relates to the primary key of table month (type: *key* and *join* node) represented in the *key* node on line 19. Lastly, the *key* node on line 22 is the primary key attribute of the table day (type: *key* and *join* node). These *join* and *key* nodes are consolidated in another *relation* node (lines 2–23; type *key* and *relation* node). This relation defines the complete dimension hierarchy structure of the database layer. The uppermost *relation* node then defines the foreign key (line 24) in the fact table which references to the primary key of the day table (line 22). This structure solely represents the inner join statements of the FROM-clause in a SQL statement when navigating over the dimension hierarchy time month. Therefore, this XML structure can be directly interpreted into the FROM part of an SQL statement as represented in Listing 5.6.

```
<relation>
  <relation>
    <join>
      <relation>
        <table>day</table>
        <key>fk_m</key>
      </relation>
      <relation>
        <join>
          <relation>
            <table>month</table>
            <key>fk_y</key>
          </relation>
          <relation>
            <table>year</table>
            <key>y_id</key>
          </relation>
        </join>
        <key>m_id</key>
      </relation>
    </join>
    <key>d_id</key>
  </relation>
  <key>fk_d</key>
</relation>
```

Listing 5.5 Hierarchy time month of dimension time

```
FROM fact
  join day on fact.fk_d = day.d_id
  join month on day.fk_m = month.m_id
  join year on month.fk_y = year.y_id
```

Listing 5.6 From clause of SQL Statement based on XML relation tree

Next to the *relation* node, the node *level* describes each hierarchy level of the dimension hierarchy. In contrast to the *relation* node, which describes the table structure on the database layer, the *level* node describes how each level is presented and from which relation the corresponding information is extracted. Thus, the *level* node is composed of the nodes *name*, holding the name of the level, *relation* and optionally *bridge*. In contrast to the *relation* node explained above, the *relation* node in node *level* is a simple relation and thus only provides the sub-nodes *table*, *key* and optionally *column*. This *relation* node is used to define which table holds the dimensional attributes of the hierarchy level. The *table* node defines the corresponding table, the *key* node defines the primary key of the table and the *column* node defines which table attributes should be presented in the user interface for this

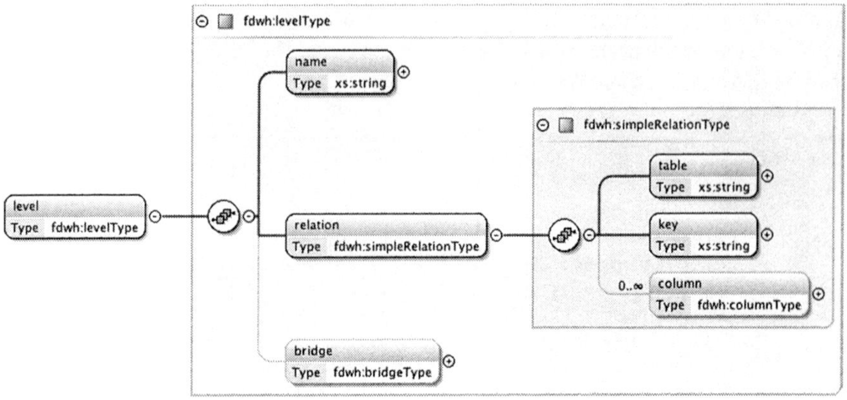

Fig. 5.7 Node "level", simple node "relation" and their children

level. Figure 5.7 illustrates the node *level*, its children and the children of simple node *relation*.

Listing 5.7 exemplifies the *level* nodes of dimension hierarchy time month. The *level* node from lines 1 to 10 describes the level day, *level* node from 11 to 21 describes level month and the last *level* node describes the level year. In the sub-node *name*, the name of the corresponding level is defined. Each *relation* node identifies which table attribute has to be displayed and the table in which the attribute is found. For level day, the corresponding table is day (line 4) and the attribute to be shown on the user interface is weekday (line 7). On the user interface it is displayed as "Week Day", which is defined in node *display* on line 8.

```
<level>
  <name>day</name>
  <relation>
    <table>day</table>
    <key>d_id</key>
    <column>
      <column>weekday</column>
      <display>Week  Day</display>
    </column>
  </relation>
</level>
<level>
  <name>month</name>
  <relation>
    <table>month</table>
    <key>m_id</key>
    <column>
      <column>monthname</column>
      <display>Month</display>
    </column>
  </relation>
</level>
```

```
<level>
  <name>year</name>
  <relation>
    <table>year</table>
    <key>y_id</key>
    <column>
      <column>year</column>
      <display>Year</display>
    </column>
  </relation>
</level>
```

Listing 5.7 Level nodes of dimension hierarchy time month

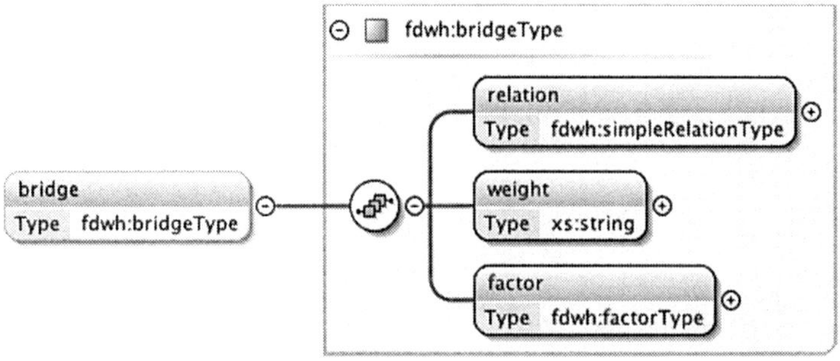

Fig. 5.8 Node "bridge" and its children

The optional *bridge* node in node *level* is used when the dimension levels are related to each other over bridge tables (see Example 17 for further information about bridge tables). This is for example the case with dimension movie. The dimension level movie is related to level producer over a bridge table thus a movie can be produced by multiple producers. In order to correctly aggregate the facts from level movie to level producer, it is necessary to know how the weighting factor of the bridge table is applied. Hence, a *bridge* node must be specified for level producer. Sub nodes of the bridge node are *relation* of type simple relation, *weight* and *factor*. The *relation* node specifies the bridge table and its foreign key to the level above the bridge table. The *weight* node specifies the attribute in the bridge table holding the weighting factor. The *factor* node specifies the operation how the weighting factor should be applied. Figure 5.8 illustrates the *bridge* node and its children.

Listing 5.8 shows the *level* nodes for the hierarchy producer of dimension movie. The level producer contains the *bridge* node (lines 42–49) in order to define how facts are aggregated from movie to producer. Its weighting factor is defined in the node *weight* on line 47 and the node *factor* on line 48 specifies the mathematical operation PROD. When a fact is aggregated to producer, the weighting factor is multiplied with the aggregated fact in order to give the final result on dimension level producer.

```
<level>
  <name>movie</name>
  <relation>
    <table>movie</table>
    <key>m_id</key>
    <column>
      <column>name</column>
      <display>Name</display>
    </column>
    <column>
      <column>overview</column>
      <display>Overview</display>
    </column>
    <column>
      <column>director</column>
      <display>Director</display>
    </column>
    <column>
      <column>actors</column>
      <display>Actors</display>
    </column>
    <column>
      <column>release_date</column>
      <display>Released</display>
    </column>
    <column>
      <column>runtime</column>
      <display>Runtime</display>
    </column>
  </relation>
</level>
<level>
  <name>Producer</name>
  <relation>
    <table>producer</table>
    <key>p_id</key>
    <column>
      <column>studio</column>
      <display>Producer Studio</display>
    </column>
  </relation>
  <bridge>
    <relation>
      <table>mov_cat_bridge</table>
      <key>p_id</key>
    </relation>
    <weight>weight</weight>
    <factor>PROD</factor>
  </bridge>
</level>
```

Listing 5.8 Level nodes for dimension hierarchy movie producer including bridge nodes

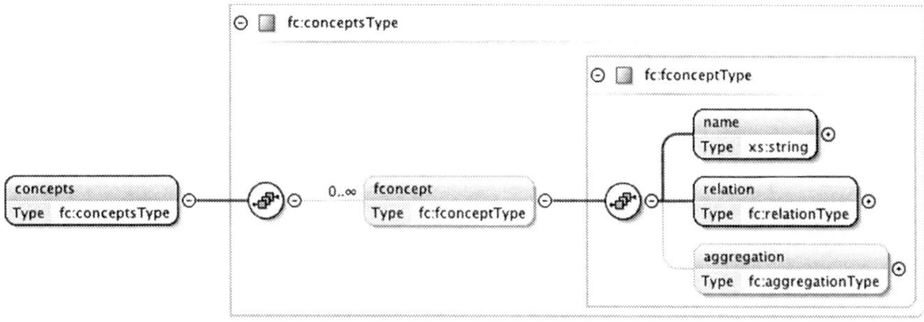

Fig. 5.9 Root node "concepts", node "fconcept" and its children

5.3.2 XML Schema for Fuzzy Concepts

Based on the discussion in Sect. 3.4.4, a XML schema for fuzzy concepts can be specified. The XML schema describing the fuzzy concept consists of the root node *concepts*. The *concepts* node can have zero or more *fconcept* nodes that hold information about the fuzzy concepts. It is worth noting that the node *concepts* might be empty when no fuzzy concept is specified or when the user removes all fuzzy concepts within the fuzzy concept administration user interface. This behavior is not fully compliant with the fuzzy concept meta model in Sect. 3.2.4 as the meta model defines a fuzzy data warehouse having at least one fuzzy concept. However, it might be possible to remove all fuzzy concepts and turn the fuzzy data warehouse into a crisp data warehouse. Consequently the prototype should implement this possibility.

The node *fconcept* is composed of the nodes *name*, *relation* and optionally the node *aggregation*. The *name* node contains the name of the specified fuzzy concept. The relations of the tables in the database are specified within the *relations* node, which is discussed in greater detail below. When the fuzzy concept can be aggregated by propagation, aggregation or applied to volatile cubes, the *aggregation* node is specified. Figure 5.9 illustrates the root node *concepts*, its child node *fconcept* and its children.

The *relation* node holds the information about the fuzzy concept structure in the database layer. It consists of the five child nodes *ta*, *fmt*, *fct*, *memfunc* and *trigger*. The node *ta* provides the information about the target attribute of the fuzzy concept. Therefore, it is further divided into the nodes *table*, *key* and *target*. The *table* node defines the table in which the target attribute resides and the *key* node defines the primary key of the table. Finally, the *target* node defines the target attribute. Either the target attribute is an attribute of the table, or for volatile target attributes, the *target* node defines the SQL query how the target attribute is aggregated. Listing 5.9 illustrates the *ta* node of a volatile target attribute. The *ta* node specifies the target attribute for the fuzzy concept city store surface. The target attribute is aggregated from the attribute store surface that resides in the hierarchy level store. Consequently, the target attribute table is city, as the fuzzy concept is applied to level city, but the target attribute itself has to be aggregated first by a SQL query.

```
<ta>
  <table>city</table>
  <key>c_id</key>
  <target>
    <query>select c_id, sum(surface) as ta from fdwh.store join
        fdwh.city on store.fk_c = city.c_id group by c_id</query
        >
  </target>
</ta>
```

Listing 5.9 TA node of fuzzy concept city store surface

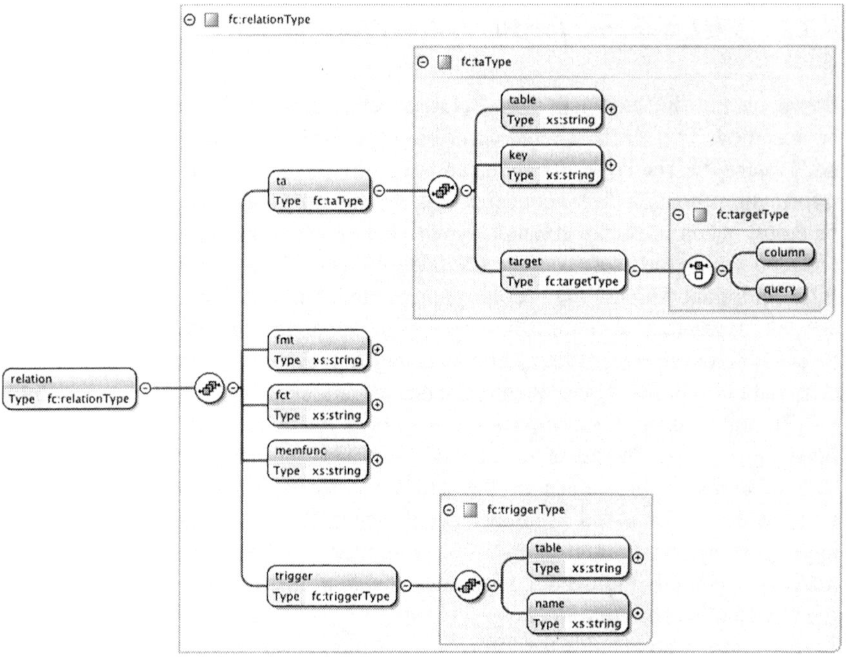

Fig. 5.10 Node "relation" and the complete structure of its children

The nodes *fmt*, *fct* and *memfunc* are simple nodes that define the name of the fuzzy membership table, fuzzy classification table and the stored procedure holding the membership function. In contrast to crisp dimension and fact definitions, these tables are precisely defined in their structure and therefore no additional information, such as primary key attribute, is needed. As described in Sect. 5.1, the membership functions are directly integrated as stored procedures into the database layer. Consequently, the XML schema does not have to provide more information about the membership function than the name of the stored procedure. The fuzzy concept administration engine can directly access and modify the stored procedure in the database tier. The *trigger* node is more complex. For aggregated fuzzy concepts, the trigger has to be applied to the table where the target attribute originates from and not to the table specified in the *ta* node. Hence the node *trigger* is composed of a node *table* and a node *name*. The *table* node contains the name of the table on which the trigger is applied and the *name* node defines the name of the trigger. Figure 5.10 shows the complete structure of the node *relation* and its children.

Listing 5.10 illustrates the *relation* node of the fuzzy concept city store surface. Lines 2–8 describe the *ta* node that is previously discussed in Listing 5.9. The nodes *fmt*, *fct* and *memfunc* contain the name of the corresponding tables and stored procedure (lines 9–11). It is interesting to note that *memfunc* not only contains the name of the stored procedure but also a parameter 'city'. The fuzzy concept administration engine can subsequently take the complete string from *memfunc* and pass it directly into an SQL statement without previously parsing it. Finally, the *trigger* node defines the table store and the trigger function fc_storesurf_city_trigger on line 12–15.

```
<relation>
  <ta>
    <table>city</table>
    <key>c_id</key>
    <target>
      <query>select c_id, sum(surface) as ta from fdwh.store
          join fdwh.city on store.fk_c = city.c_id group by c_id
          </query>
    </target>
  </ta>
  <fmt>fmt_storesurf_city</fmt>
  <fct>fct_storesurf</fct>
  <memfunc>fc_storesurf_prop('city')</memfunc>
  <trigger>
    <table>store</table>
    <name>fc_storesurf_city_trigger</name>
  </trigger>
</relation>
```

Listing 5.10 Relation node of fuzzy concept city store surface

The last child node of *fconcept* is the optional node *aggregation*. This node is only used when the fuzzy concept is aggregated over dimensions of the fuzzy data warehouse or applied to volatile cubes. As defined in Sect. 3.4.2, a fuzzy concept can be aggregated or propagated in the fuzzy data warehouse. Hence, the *aggregation* node comprises the three sub-nodes *aggregator*, *propagation* and *volatile*. If a fuzzy concept can be aggregated, the *aggregator* node contains the aggregation operation that is used to aggregate the membership degrees. The operation can be: SUM, MAX, MIN, AVG or PROD. When propagated, the *propagation* node contains the propagated fuzzy concepts. So the node *propagation* comprises one or more *fconcept* nodes. Finally, when the fuzzy concept can be applied to volatile cubes, the node *volatile* is specified. With the child node *ta* of the node *volatile*, the target attribute, which has to exist in the volatile cube, is defined. It has to be stated here,

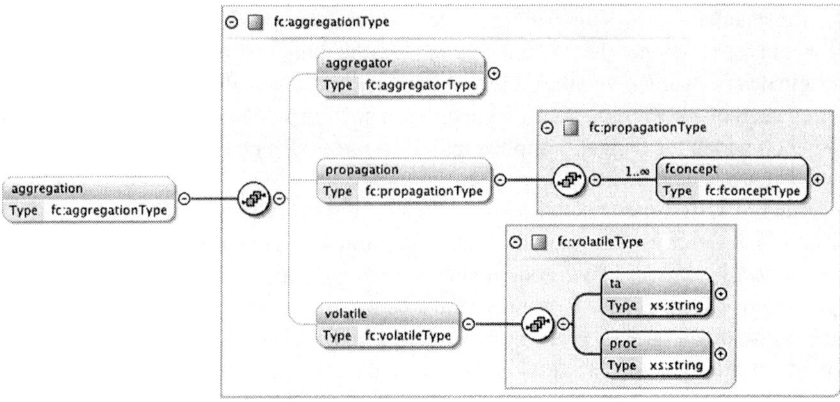

Fig. 5.11 Node "aggregation" and its children

that the *ta* node is of type simple and only contains the name of the target attribute. The child node *proc* defines the stored procedure that calculates the membership degrees of the volatile cube. This stored procedure takes as a parameter the SQL statement of the volatile cube, the target attribute and returns the volatile fuzzy cube with the membership degrees. Such a stored procedure is discussed in Sect. 3.4.5 and illustrated in Listing 3.3. Figure 5.11 visualizes the node *aggregation* and its children.

Listing 5.11 provides the complete *aggregation* node of the fuzzy concept store surface. This fuzzy concept can be propagated on the hierarchy level city and region. Therefore, for each hierarchy level a new fuzzy concept is created. In the *aggregation* node, the two fuzzy concepts are defined as *fconcept* nodes (for city lines 3–21 and for region lines 22–40). It is noteworthy that the fuzzy concepts have different fuzzy membership tables, but the fuzzy classification table remains the same as in the original fuzzy concept. The stored procedure for calculating the membership degrees in both fuzzy concepts is the same, but the calling parameter for the procedure differs (see line 15 and 34). Both fuzzy concepts define a trigger on the store table as shown on lines 16–19, respectively lines 35–38.

```
<aggregation>
   <propagation>
    <fconcept>
     <name>City Store Surface</name>
     <relation>
      <ta>
       <table>city</table>
       <key>c_id</key>
       <target>
        <query>select c_id, sum(surface) as ta from fdwh.store
            join fdwh.city on store.fk_c = city.c_id group by
            c_id</query>
       </target>
```

```
        </ta>
        <fmt>fmt_storesurf_city</fmt>
        <fct>fct_storesurf</fct>
        <memfunc>fc_storesurf_prop('city')</memfunc>
        <trigger>
          <table>store</table>
          <name>fc_storesurf_city_trigger</name>
        </trigger>
      </relation>
    </fconcept>
    <fconcept>
      <name>Region store surface</name>
      <relation>
        <ta>
          <table>region</table>
          <key>r_id</key>
          <target>
            <query>select r_id, sum(surface) as ta from fdwh.store
                join fdwh.city on store.fk_c = city.c_id join fdwh.
                region on city.fk_r = region.r_id group by r_id</
                query>
          </target>
        </ta>
        <fmt>fmt_storesurf_region</fmt>
        <fct>fct_storesurf</fct>
        <memfunc>fct_storesurf_prop('region')</memfunc>
        <trigger>
          <table>store</table>
          <name>fc_storesurf_region_trigger</name>
        </trigger>
      </relation>
    </fconcept>
  </propagation>
</aggregation>
```

Listing 5.11 Aggregation node of fuzzy concept store surface

According to the discussion in Sect. 3.4.4, fuzzy concepts on facts often contains aggregations over volatile cubes. The fuzzy concept user rating is a concept which defines aggregation using either aggregation of the membership degrees with an average operation or over volatile cubes. In Listing 5.12, the *aggregation* node of the fuzzy concept user rating is illustrated. The *volatile* node (lines 3–6) defines how the fuzzy concept might be applied to volatile cubes. The volatile cube must contain the target attribute user_rating. If this is the case, the stored procedure fc_rating_vol can be used to calculate the membership degrees. It is subsequently called with the SQL statement of the cube as first parameter and the target attribute name as second.

```
<aggregation>
  <aggregator>AVG</aggregator>
  <volatile>
    <ta>user_rating</ta>
    <proc>fc_rating_vol</proc>
```

```
  </volatile>
</aggregation>
```

Listing 5.12 Aggregation node of fuzzy concept user rating

5.3.3 *The Query and the Fuzzy Concept Administration Engine*

The query engine is realized within the thesis of Stefan Nüesch [Nüe11]. In [Nüe11], an OLAP cube implementation is presented which provides the crisp part of the business logic and the visualization of the fuzzy data warehouse. The fuzzy part, including the fuzzy cube visualization and the fuzzy concept administration engine, is discussed within the thesis of Christoph Rathgeb [Rat11].

The query and administration engines provide functionalities for parsing, editing the XML files and querying the database. The parsing and editing functionalities for the XML files are realized with the Java Architecture for XML Binding (JAXB) [Inc11b]. JAXB allows the creation of a Java object for every node specified in an XML schema definition. The content of an XML file, that follows the schema definition, can then be loaded and edited in the Java objects. This method allows efficient handling of XML files. Further, the administration engine can efficiently edit the fuzzy concept meta information by Java object rather than by editing XML files.

Communication from the engines to the database is performed with the JDBCTemplate class of the Spring Framework [Inc11e]. The JDBCTemplate reduces the complexity of low level communications with databases such as opening, closing connections and parsing database exceptions [Inc11d]. With the JDBCTemplate, the engines have a persistence tier for communication with the underlaying database. Similar to JAXB, it projects the relational data into Java objects. This function is commonly known as object-relational mapping [Gon09].

The query engine provides information about the fuzzy data warehouse structure and transforms request from the visualization layer into database queries. The structure is defined in the XML files that are read out during the startup process of the fuzzy data warehouse. It then passes the structural information to the visualization layer in order to create the visualization elements of the fuzzy data warehouse. When the visualization layer passes a request to the query engine, it transform the request into a database query using the JDBCTemplate. Once the database returns a query result, it is prepared for the visualization layer. The preparation task distinguishes whether the visualization layer requested a pivot table or a simple table. In the case of a pivot table, the query engine parses the result and wraps it into a data table object. This object is then forwarded to the visualization layer. For simple table, the query result is directly forwarded to the visualization layer.

In contrast to the query engine, the fuzzy concept administration engine also provides write capabilities to the XML files and the database. This is the main reason the two engine are separated in the fuzzy data warehouse architecture. When the fuzzy data warehouse starts up, the administration engine reads the fuzzy concept definitions from the XML file. For a visualization request from the visualization layer, the administration engine first checks the selected crisp data warehouse elements and then returns the valid fuzzy concepts. A fuzzy concept is only valid if it can be applied to one of the selected crisp data warehouse elements. This restriction simplifies navigation of the fuzzy data warehouse because the end user only sees valid fuzzy concepts for his query. The fuzzy query results are processed by the query engine and further passed to the visualization layer. The second task of the administration engine is the management of the fuzzy concepts. This includes creating, editing and deleting fuzzy concepts. When adding a new fuzzy concept, the visualization layer passes it as an object to the administration engine. The engine then persistently stores it in the XML file by using JAXB to create a new node in the XML file. Further, the necessary database objects such as fuzzy membership table, fuzzy classification table, membership function are created. Finally, the trigger functions are fired on the database to create the initial set of membership degrees. When editing a fuzzy concept, the objects of the fuzzy concept is manipulated and then persistently stored. The deletion operation deletes the fuzzy concept objects and all references to them in the XML file and in the database.

5.4 Visualization

The visualization layer provides the user interface for the fuzzy data warehouse, allows navigation of the fuzzy data warehouse and administration of fuzzy concepts. Stefan Nüesch [Nüe11] and Christoph Rathgeb [Rat11] contributed the visualization layer for the fuzzy data warehouse.

To create the visualization layer, Java Server Faces [Inc11c] and ICEFaces [INC11a] are used. Java Server Faces is the standard framework for user interfaces in Java Enterprise Edition [Hef07]. It is based on Java Server Pages and Enterprise JavaBeans [Hef11] and allows simple development of the visualization components in web application. ICEFaces is a framework that enables AJAX [Per06, GbGA06] for Java Server Faces. With AJAX, requests from the visualization layer to the business layer can be handled asynchronously. In traditional web applications, the user interface waits for the answer from the business layer and is blocked as long the business layer has not finished. Using AJAX, the visualization layer processes requests to the business layer in JavaScript and therefore can continue interacting with the user while the request is on hold.

The visualization layer is divided into two parts: navigation and the fuzzy concept administration. In the navigation part, the user can query the fuzzy data warehouse

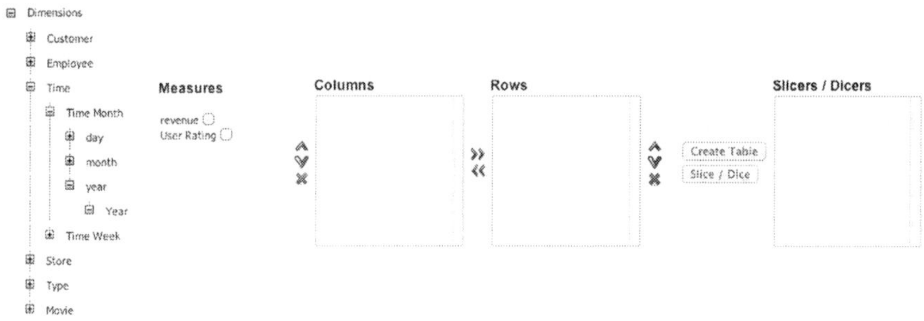

Fig. 5.12 Navigation of a sharp cube

either crisp or fuzzy. In the administration part, the user can manage fuzzy concepts in a six step wizard. The next subsections discuss the two parts in greater detail.

5.4.1 *Navigation of the Fuzzy Data Warehouse*

The navigation is split into two parts: pivot navigation and fuzzy navigation. The pivot navigation allows navigation of the fuzzy data warehouse in a sharp manner and presents the query results in a pivot table. Whereas, the fuzzy navigation allows querying of the fuzzy data warehouse fuzzily and the result set is presented in simple table form.

For both navigation forms, the user has similar user interfaces. Figure 5.12 presents the navigation elements of a pivot navigation. On the left-hand side, a tree with dimension elements is presented. Dimension elements can be dragged into the column or row box for selection. When a category attribute of a dimension is chosen, every dimensional attribute that belongs to the category attribute is inserted in the selection box. The selection boxes specify where in the pivot table the element is going to appear; either in the column or in the row. The facts are listed as checkboxes next to the dimension tree. Once the user selects the elements, the button "Create Table" becomes active and the user can execute a query on the fuzzy data warehouse.

The selected dimension elements can be moved from one box to another by using the double arrows between the boxes. Further, the position of the elements can be changed or removed with the arrows and the cross at the side of the boxes. Figure 5.13 shows a query of the fuzzy data warehouse. The years in the column box, the region in the row box and the fact revenue are selected. The pivot table of the query result is presented below the navigation.

For slicing and dicing, the user can define the slicers by clicking on the "Slice/Dice" button. A popup window opens, in which slicers can be defined based on the selected dimension elements. Figure 5.14 shows the popup window with two slicers defined: Year EQUALS 2010 and Region EQUALS German. The slicers

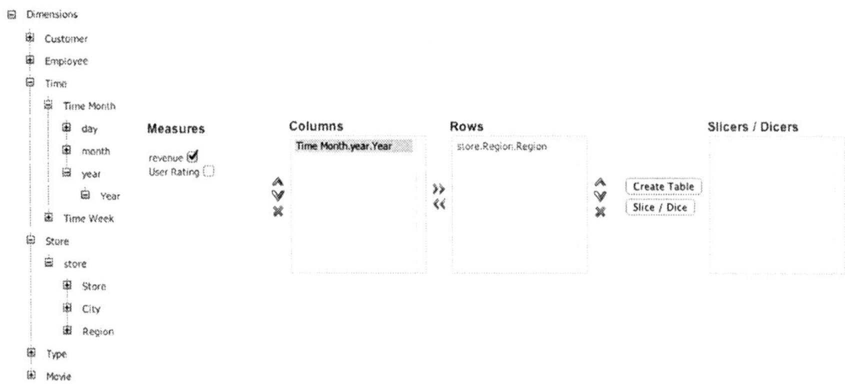

	year	Year	2005	2006	2007	2008	2009	2010	2011
Region									
Region									
French			3420129	3378420	3420412	3435903	3453435	3434462	382271
German			3985124	3872219	3907133	3937668	3942517	3942741	445704
Italian			267810	269101	272069	248559	257313	266034	29125
Rhaeto-Romanic			270619	261144	263823	255155	262244	275487	26956

Fig. 5.13 Navigation of the fuzzy data warehouse including a pivot table

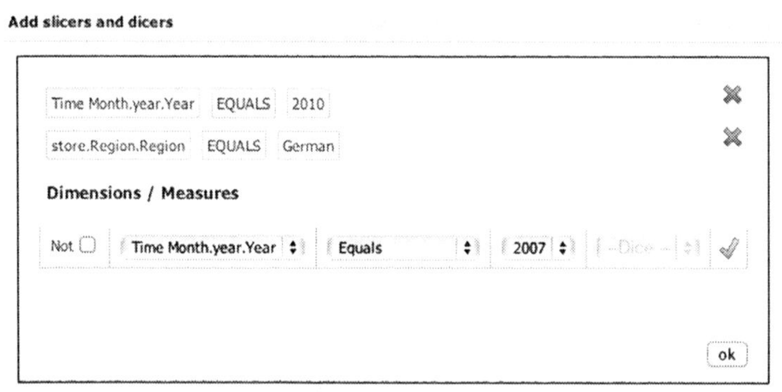

Fig. 5.14 Slice/Dice window with two selected Slicers

are further displayed in the third selection box called "Slices/Dices". Figure 5.15 presents the navigation and the resulting pivot table of the query with the slicers.

The navigation elements of the fuzzy navigation are similar to the elements of the pivot navigation. Figure 5.16 shows the fuzzy navigation elements. The tree on the left hand side contains not only dimension elements, but also fuzzy concept elements. The shown fuzzy concepts depend on the selection of the fact and the dimension elements. In Fig. 5.16, the fuzzy concept store revenue, store surface and monthly revenue are shown. These fuzzy concepts are the only concepts that can be applied to the selected fact revenue and the selected dimensional attributes month

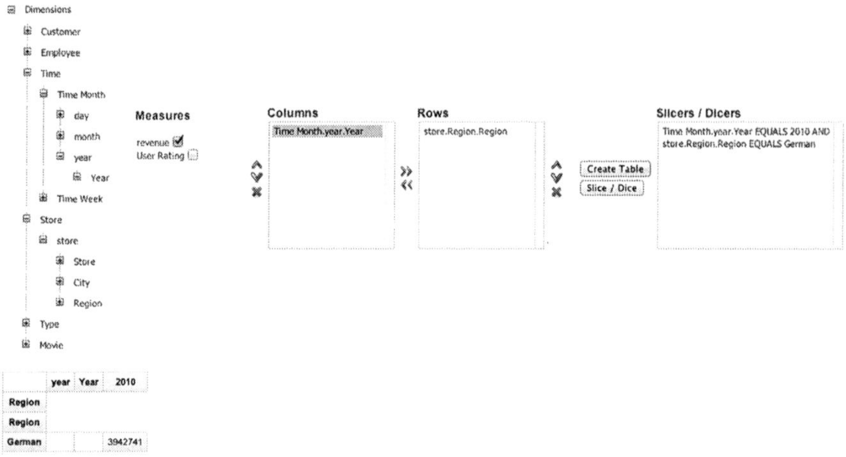

Fig. 5.15 Navigation of the fuzzy data warehouse including Slicers and a pivot table

Fig. 5.16 Navigation of a fuzzy cube

and city. The selection boxes are now labeled "Dimension Attributes" and "Fuzzy Concepts". In the first box, the dimensional attributes can be dragged in, and in the second box the fuzzy concepts can be dragged in. The first box is the conjunction of the two boxes in the pivot navigation. A selection in the pivot navigation is automatically applied to the fuzzy navigation. Therefore, a user can first query a pivot table and then switch to the fuzzy navigation, preserving the selection of the first query.

In the tree navigation, the fuzzy concepts are listed similarly to the dimensions. First, the fuzzy concept name is shown. Beneath the name, the linguistic terms are listed. It is possible to drag the fuzzy concept name to the fuzzy concept box. That way, all linguistic terms are dragged to the box. Figure 5.17 shows the navigation with the fuzzy concept monthly revenue and the resulting cube. In the

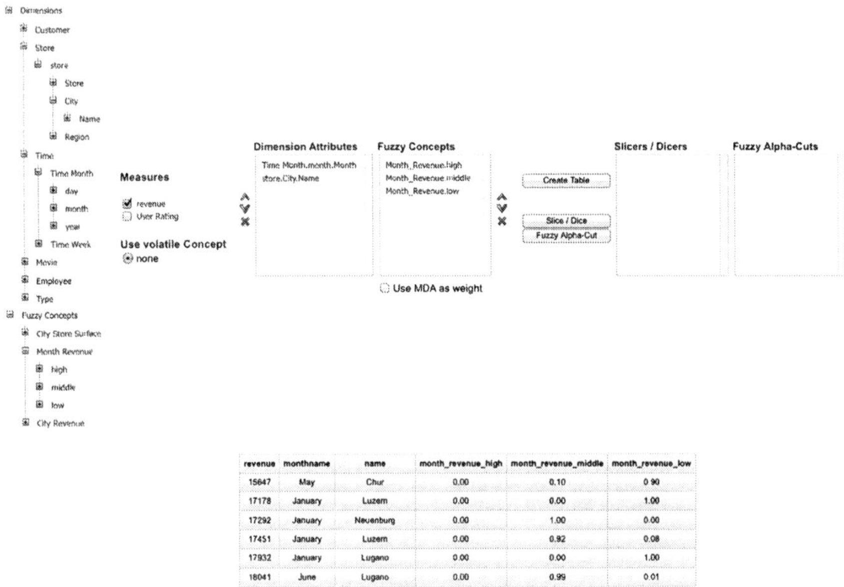

Fig. 5.17 Fuzzy navigation of the fuzzy data warehouse

fuzzy navigation, the slicing window is extended with the possibility to do α-cuts. α-cuts can be selected in the slice window the same way as sharp slices. Furthermore, it is possible to select volatile fuzzy concepts or to use the membership degree attributes for the aggregation of the facts.

5.4.2 Administration of Fuzzy Concepts

Administrating fuzzy concepts can be done in a six-step wizard. In the first step, the user can chose a fuzzy concept to either edit or delete. Additionally, a new fuzzy concept can be added by clicking on the "add" link. In the second step, the name of the fuzzy concept, the table in which the target attribute resides and the target attribute have to be set. The target attribute can either be a column in the target attribute table or a query that returns the target attribute and its key. Figure 5.18 presents in (a) the first step in (b) the second step with a column as target attribute and in (c) the second step with a query as target attribute.

Linguistic terms can be managed in the third step. Next, the membership functions have to be defined. First, the type of fuzzy concept is specified. Then, the user has the choice of three types of functions: trapezoid, discrete and manual. Different options are presented based on the type of the membership function. The manual membership function does not provide any options because it assumes

Fig. 5.18 First and second step of the wizard

that the user defines the membership degrees directly in the fuzzy membership table. Additionally, it has to be specified on which table the trigger, that calls the membership function, has to be applied. Figure 5.19 shows (a) the third step, (b) the fourth step with the options for a trapezoid membership function and (c) the fourth step with the options for a discrete membership function.

In the fifth step, how the fuzzy concept can be aggregated, propagated or if it can be a volatile fuzzy concept must be specified. For aggregation, an aggregation operation can be chosen. When the fuzzy concept can be propagated, the siblings can either be chosen from the existing fuzzy concepts or new fuzzy concepts can be defined. When a new fuzzy concept has to be defined, the wizard restarts from step 2. In this case, it already contains information such as linguistic terms from the parent concept. If the fuzzy concept should be available as volatile fuzzy concept, the checkbox under volatile has to be activated. Figure 5.20 shows the fifth step

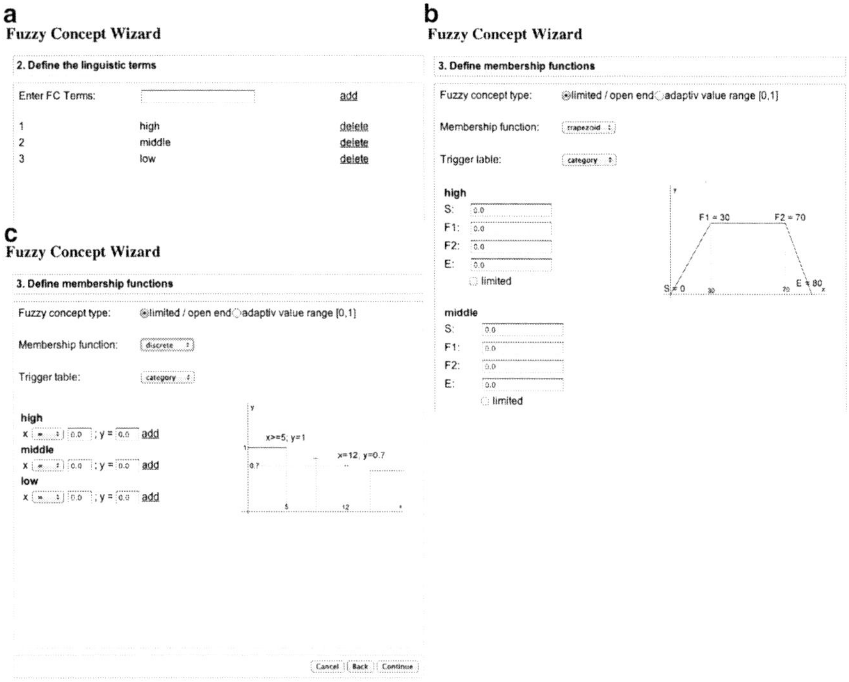

Fig. 5.19 Third and fourth step of the wizard

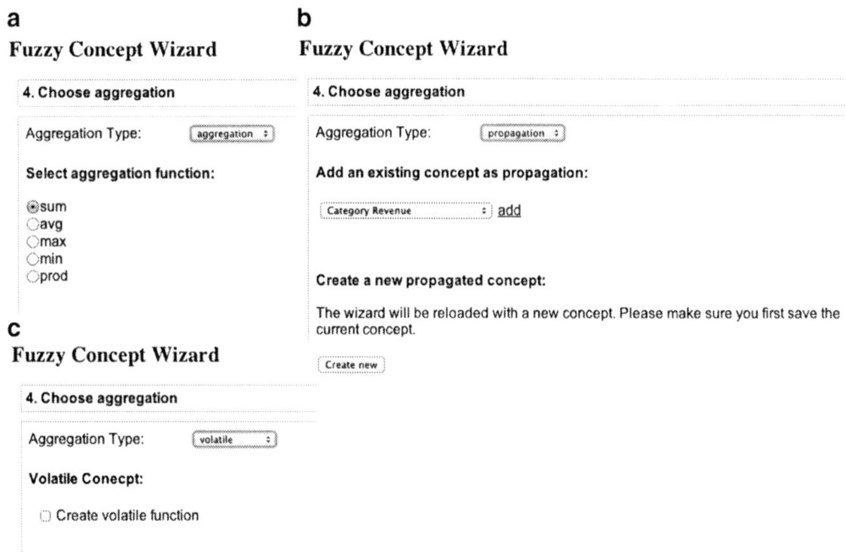

Fig. 5.20 Fifth step of the wizard

with the options for aggregation in (a), the options for propagation in (b) and the checkbox option for a volatile fuzzy concept in (c).

Finally, step 6 provides a summary of the fuzzy concept and a button to save the concept. When saving, the fuzzy concept is passed as an object to the administration engine and the engine processes the necessary task on the database and in the XML file. In every step in the wizard, the user has the possibility to go back or to cancel the process.

Part IV
Evaluation and Conclusion

Chapter 6
Evaluation and Conclusion of Fuzzy Data Warehouse

6.1 Evaluation

6.1.1 Concept

Benefits of the Concept

As shown in Chap. 4, the application of fuzzy concepts for analysis can provide better results than classical analysis. The interpretation of measures in non-numeric, meaningful linguistic terms provide a more accurate interpretation of measures for business users. In [FZ09], a fuzzy data warehouse was proposed for a web analytics system. In this research paper, it is shown that the raw measures of web analytics are not always simple to interpret. The metric page visits counts how often a user visited a web site. It is not always possible to count visits of distinct users accurately. Therefore, this metric is imprecise. When classifying users, a crisp concept of page visits is only limited accurate. A user that has 75 visits might be classified as a bad visitor, whereas a visitor with 80 visits might be classified as good visitor. Because of the imprecision of the metric, a user might be classified in the wrong class. It is therefore difficult to compare users. By using a fuzzy classification the user can belong to both linguistic terms at the same time. This allows a smoother transition between the classes and it can better handle the imprecision of the metric. The fuzzy concept page visit is therefore a more accurate classification than the crisp classification.

As discussed in Sect. 3.1, approaches to integrate fuzzy concepts in data warehouse and OLAP cubes have been already discussed in the past. However, most of these approaches have been developed for a certain field of application and they often imply restrictions on how fuzzy concepts can be applied. It has been proposed that fuzzy concepts are integrated into the hierarchy structure of dimensions. This implies the normalization of fuzzy concepts. Other propositions replace the original values of the data warehouse with information from the fuzzy concept.

D. Fasel, *Fuzzy Data Warehousing for Performance Measurement*, Fuzzy Management Methods, DOI 10.1007/978-3-319-04226-8_6,

Consequently, it is no longer possible to analyze the original values of the data warehouse. The proposed concept in Sect. 3.2 addresses these drawbacks and provides a more flexible method of integrating fuzzy concepts using a meta table structure. With meta tables, it is possible to integrate fuzzy concepts on every hierarchy level of dimensions without affecting the hierarchy path or the summarizability of the dimension. Hence, the fuzzy concepts can stay unnormalized. The fuzzy concepts might be normalized during runtime only in specific analysis operation (see Example 24 for a operation using normalization).

Another major improvement of the proposed concept is the ability to analyze the fuzzy data warehouse simultaneously crisp and fuzzy. Due to the fact that the information of the fuzzy concepts is rolled out into meta tables, the fuzzy data warehouse can be queried crisp and by applying the presented fuzzify and defuzzify functions in Sect. 3.5.2, the crisp query can be extended with the fuzzy concept information. With regard to the existing approaches, only approaches that integrate the crisp and fuzzy information into multiple cubes can offer a similar functionality, but in order to receive both information, multiple cubes have to queried.

Aggregation of the fuzzy concepts is discussed in detail and different aspects of aggregation are presented. The fuzzy concept can be aggregated, propagated or can even be applied to volatile cubes that only exist during the query process. This discussion of aggregation, compared to existing literature, is much more detailed and allows handling fuzzy concepts very similarly to facts or dimensional attributes. Hence, it improves the integration of fuzzy concepts in data warehouses.

A functionality that has not yet been proposed is the possibility to create variants of the fuzzy concepts. Because of the separation of the linguistic term and membership degrees in two meta tables, it is possible to combine these tables in different variations with each other. For instance, a fuzzy concept and its variant can classify a target attribute with the same membership degrees but different linguistic terms. Vice versa, another fuzzy concept and its variant have the same linguistic terms but use other membership functions to calculate the membership degrees of the target attributes. For example, the fuzzy concept user rating discussed in Sect. 4.2.6 might be not only used for creating the list of movie hits but also for internal decision making about how prominently a movie should be placed in the store. The fuzzy concept can then be extended with a variant having other class membership attributes such as "showcase", "front store", "backend". This functionality further improves the flexibility of the concept proposed in this thesis.

A final benefit of the proposed concept, with regard to the discussed approaches in Sect. 3.1, is the ability to apply fuzzy concepts in the modeling process of a data warehouse as well as on already existing classical data warehouses. To better support application on existing data warehouses, a method for modeling a fuzzy data warehouse is presented in Sect. 3.3. The meta table structure of the fuzzy data warehouse allows the extension of operational data warehouses with fuzzy concepts without having to touch the existing data structure of the data warehouse.

To sum up, it can be stated that the proposed concept in Sect. 3.2 can be used to implement all approaches discussed in Sect. 3.1. Furthermore, it provides additional functions such as fuzzy concept variants, simultaneous analysis of fuzzy and crisp data, application on new and operational data warehouses and highly flexible aggregation of fuzzy concepts. In comparison to classical data warehouses and fuzzy data warehouse approaches shown in Sect. 3.1, the flexibility and the set of new functionalities of the proposed concept can improve the analysis of data considerably.

Limitations of the Concept

The fuzzy data warehouse concept in Sect. 3.2 is based on a relational approach for data warehouses. Therefore, it uses snowflake schemas for the classical data warehouse part. The meta model in Sect. 3.2.4 further substantiates the relational approach by using fact table and dimension table classes in the classical data warehouse class. Relational databases, especially when the data is stored in third normal form, can negatively influence the performance of data warehouses. For data processing in OLAP systems, specific multidimensional database applications such as Hyperion [Ora10] have been developed. Hyperion stores data in blocks that are arranged according to the dimensions. By this type of storage engines, facts can be modeled in a separate measure dimension similar to the multidimensional cube described by Agrawal et al. in [AGS95]. Therefore, the meta model described in Sect. 3.2.4 has to be adapted before being applied to multidimensional OLAP cubes. It is necessary to analyze in future research whether the meta table structure can be implemented in multidimensional databases or if a new implementation of fuzzy concepts would be necessary.

One of the main characteristics of a data warehouse is that data is stored on a timely basis. Therefore, a time dimension is mandatory in order to have data in different states at different points of time (e.g.: Revenue of Store A in 2011, Revenue of Store A in 2012). The meta table structure has not foreseen any chronological versioning or historization of fuzzy concepts. Subsequently, it is not possible to have different versions of fuzzy concepts in different points in time. The ability to see how data was classified at a certain time might be of interest. In order to this, the meta table structure would have to be extended with temporal attributes as proposed for classical data warehouse structures [CS99, Ede01, MV00].

Finally, the fuzzy data warehouse concept describes how to implement fuzzy concepts but does not explain how business users can derive such concepts from real world situations. The end user still has to define manually the linguistic terms and the membership function of the fuzzy concepts. No method is provided to derive the membership functions or degrees from data or from external sources. In order to derive fuzzy concepts from data, approaches like inductive fuzzy classification [Kau12] has been analyzed. However in this thesis, this topic has been considered as out of scope.

6.1.2 Application and Implementation

Benefit of the Application and Implementation

Chapter 4 illustrates how a fuzzy data warehouse can be used to improve analysis of a virtual movie rental company. Different analyses involving fuzzy concepts are executed in order to show the benefit of the fuzzy concepts. Example 28 uses three fuzzy concepts in a single analysis. Two concepts are propagated fuzzy concepts that show classifications of category attributes. The third concept is a volatile fuzzy concept that classifies the cube measures. With a single analysis, the movie rental company can not only see the sales performance of movies per month but also how strong monthly sales and overall movie sales are. In order to achieve such an analysis with a classical data warehouse, the company would have to execute at least three query statements and combine the results. Thus, the application in Sect. 4.3 provides scenarios that show the improvement in analysis using fuzzy concepts.

The implementation in Chap. 5 provides a proof of concept prototype that let endusers easily access the fuzzy data warehouse. It abstracts the database layer and uses meta information in order to create a tree structure based navigation through the data warehouse and the fuzzy concept. The meta information is structured using XML documents. The structure itself is defined with XML schema documents. As a result, the prototype can be applied to different data warehouses, which might store its data in different table structures. This flexibility allows extending the prototype in the future to support other data warehouse applications.

The fuzzy concepts are administrated using a wizard that helps the enduser through the complex process of defining fuzzy concepts. Table structures, stored procedures and triggers on the database layer are automatically created by the administration engine. This separation of data layer and meta layer improves the ability to use different database technologies and different data structure models. Yet the XML schema provides all necessary elements and dependencies for defining a complete fuzzy data warehouse as defined in Chap. 3.

Limitations of the Application and Implementation

The major limitation of the application is the fact that the movie rental company example is not based on a real world scenario. An application on a productive data warehouse would increase the validity of the fuzzy data warehouse approach presented. Nevertheless, the fuzzy data warehouse of the movie rental company uses more than 3 million fact entries from more than 5 years in order to provide a large enough data sample for analysis. The information about the movies is extracted from the open movie database [TMD10] in order to provide accurate movie data. The fact entries have been created using random generators in conjunction with information about real world movie sales and ratings from the open movie database and the internet movie database [IMD11].

The prototype is considered to be a proof of concept and only implements the navigation over the fuzzy data warehouse and the administration of the fuzzy concepts. In order to provide a fully operational data warehouse, the prototype has to be extended with support for user management and with ETL functionalities. Furthermore, caching of the data in memory would greatly increase the performance for querying the fuzzy data warehouse.

The data structure of the crisp data warehouse part is modeled using a snowflake schema. Subsequently, dimensions are built with tables in the third normal form. When querying the fuzzy data warehouse, cost intensive inner join statements have to be used. The database might perform better when using a data model that does not imply third normal form such as the star schema. Another possibility might be the creation of aggregated views that denormalize the dimension structure and provide flat views on dimensions [CS96]. Furthermore, another database engine such as Vertica [Ver11] might be used. Vertica stores data in a column-based access method rather than a traditional row-based access method. Column based access methods can improve query performance of a data warehouse.

6.1.3 Further Outlook

It might be interesting to extend the fuzzy data warehouse concept for use with multidimensional databases such as Hyperion [Ora10]. To accomplish this, the meta table structure has to be adapted and it would be necessary to analyze whether it would make more sense to create a fuzzy concept dimension. The calculation of the membership degrees could be realized in calculation scripts (see [Ora11] for more information about calculation in Hyperion).

A further future task is the application of fuzzy concept on an operational data warehouse. The application described in Chap. 4 shows the advantage of using fuzzy concept, but is based on a virtual company. Applying the fuzzy concept to real world data would further strengthen the validity of the fuzzy data warehouse concept.

Besides the application on a productive data warehouse, the prototype should be further developed. In order to use the prototype as a productive fuzzy data warehouse, missing features such as user management, ETL functionalities and stability improvements have to be included. It might be interesting to apply the fuzzy data warehouse to different database technologies.

In an information system architecture of a company, the data warehouse is the analytical part. It might be interesting to analyze how fuzzy concepts can be applied to other information systems. Fuzzy concepts could then be maintained by operational systems and imported to the data warehouse by using the ETL process. Consequently, it might be analyzed how such ETL processes for fuzzy concepts have to be designed. Applying fuzzy concepts to the complete information system infrastructure will have further impact on the company's ability to analyze data. Therefore, the impact of decision making should be further analyzed when applying fuzzy concept on OLTP and OLAP systems of a company.

6.2 Conclusion

Data warehouse is used for analysis of businesses performance. One potential pitfall of the classical data warehouse is that the numeric values of a data warehouse may be difficult to interpret for business users, or may be interpreted incorrectly. For more accurate understanding of numeric values, business users require an interpretation in meaningful, non-numeric terms. However, if the classification between linguistic terms is crisp, true values cannot be measured and smooth transition between classes cannot take place. A solution to this is the use of a fuzzy based approach.

In Chap. 3 of this thesis, a fuzzy data warehouse modeling approach, which allows integration of fuzzy concepts as meta tables without affecting the core of a classical data warehouse is discussed. The essence of this approach is that meta table structure is added for classification, which enables integration of fuzzy concepts in dimensions and facts, while preserving the crisp data structures. The key benefits of this approach are as follows:

- Data can be analyzed in crisp and fuzzy manner.
- The model is generic enough to cover all the approaches discussed in Sect. 3.1.
- The model can be used for modeling new data warehouses or applying to an existing data warehouse.
- Aggregation of fuzzy concepts allows applying fuzzy concept to different hierarchy levels and on volatile cubes.

The proposed fuzzy data warehouse addresses the research questions that have been listed in Sect. 1.2 as described in the next paragraphs.

6.2.1 How Can Facts in Data Warehouses Be Handled Fuzzily?

The proposed fuzzy data warehouse concept does not differentiate between facts or dimensions as target attribute for a fuzzy concept. Consequently, the facts can be enriched with fuzzy concepts by taking facts as target attributes. The fuzzy membership table provides a relation to the fact table. Section 3.2 discusses in detail how the meta table structure can be created in order to build fuzzy concepts on facts.

6.2.2 How Can Dimensions in Data Warehouses Be Handled Fuzzily?

Fuzzy concepts can be applied to dimensional attributes in every level of a dimension hierarchy. By using the meta table structure, the dimension navigation path is not influenced by the fuzzy concepts and the fuzzy concepts do not have to be normalized. The fuzzy membership table provides the relation to the dimension table in which the dimensional attribute resides. Section 3.2 provides a detailed discussion of how to implement fuzzy concepts on dimensions.

6.2.3 How Can Data Warehouse Operations Like Roll-Up, Drill-Down, Slice and Dice Be Applied to Fuzzy Data in a Data Warehouse?

In Sect. 3.5, the operation roll-up, drill-down, slice and dice are defined in detail for a fuzzy data warehouse. It is notable that these operations can be applied to both fuzzy and crisp cubes. Therefore, the proposed operations can handle both fuzzy and crisp data at the same time. In addition to the definition of the classical operation, two new operation have been proposed that can turn a crisp cube into a fuzzy cube and vice versa. These operations further improve the flexibility of the fuzzy data warehouse as they allow one to switch easily from crisp to fuzzy data analysis.

6.2.4 How Can Fuzzy Data Be Aggregated?

The aggregation of fuzzy concepts over dimensions is discussed in Sect. 3.4. Briefly, three different types of aggregation of fuzzy concepts have been identified: aggregation, propagation, and application on volatile cubes. Aggregation allows aggregating the membership degrees of a fuzzy concept to a new hierarchy level. This method only provides consistent results if the crisp aggregation of the target attribute is not a summation function and does not change the value range of other hierarchy levels. An application of an aggregation has been shown with the fuzzy concept user rating in Sect. 4.2.6.

Propagation redefines the structure of the fuzzy concept on other hierarchy levels. Therefore, at a minimum the fuzzy membership table and the membership function are redefined for the new target attribute. Propagation is a flexible way to aggregate fuzzy concepts, but increases the number of meta tables in the fuzzy data warehouse.

Finally, fuzzy concepts on facts might be applied to resulting cubes of queries. These cubes exist only during the time of the analysis and the aggregated fact is not persistently stored in the fuzzy data warehouse. Fuzzy concepts can be applied to these volatile cubes by calculating the membership degrees of the non-persistent target attribute during the analysis. The end user can apply fuzzy concepts on their temporary query results which greatly improves the flexibility of the fuzzy data warehouse. As disadvantage, the calculation of membership degrees during querying of the fuzzy data warehouse increases the query time.

6.2.5 Is It Possible to Propose Generic Modeling Methods for Fuzzy Data Warehouses?

As discussed in Sect. 6.1.1, every application presented in the literature review in Sect. 3.1 can be realized with the proposed fuzzy data warehouse concept. The problems identified in the discussed approaches can be omitted by outsourcing the fuzzy concepts into meta tables. The meta model of the fuzzy data warehouse takes a common snowflake modeling approach into consideration. It is therefore possible to integrate the fuzzy data warehouse model into the conventional data warehouse modeling process. It can be stated that the proposed fuzzy data warehouse approach is able to provide a generic modeling approach. It fits for modeling existing fuzzy data warehouse approaches and furthermore smoothly integrates into classical data warehouse modeling. However, the snowflake schema is optimized for data warehouse on relational database technologies. It has to be noted that the fuzzy data warehouse model is optimized for relational OLAP applications. When using multidimensional OLAP applications, the fuzzy data warehouse model has to be adapted.

6.2.6 How Is It Possible to Integrate a Fuzzy Data Warehouse into an Existing Information System Infrastructure in a Company in the Easiest Manner?

The meta table structure separates the fuzzy concepts from the actual data warehouse structure. This is also pointed out in the meta model in Sect. 3.2, in which the fuzzy data warehouse model is separated from the classical data warehouse model (see Fig. 3.2). Therefore, existing classical data warehouses can be transformed into fuzzy data warehouses by adding the meta table layer to the data warehouses. This will not impact the structure or data of the classical data warehouse. Furthermore, Sect. 3.3 describes a method for modeling a fuzzy data warehouse based on a classical data warehouse. This method aims to provide users with a guideline on how to transform classical data warehouses into fuzzy data warehouses.

6.2.7 What Kind of Technologies Can Be Used for Implementing a Fuzzy Data Warehouse?

In order to answer this question, a prototype was implemented in Chap. 5. The implementation is inspired from the open source data warehouse Mondrian from Pentaho Corporation [Pen10]. The technology used for the database layer is PostgreSQL [Pos11a], which is a commonly used open source relational database management system. Further, the meta information that is used for transposing the dimension structure, the facts and the fuzzy concepts onto the table structure is implemented by XML documents [HM05]. XML schema documents have been used to define the structure and the necessary elements in the XML documents. The engines for the query translation, the fuzzy concept administration and the visualization have been realized using Java. The prototype shows that the fuzzy data warehouse can be realized with common technologies that have already been used, for instance Mondrian, for implementing classical data warehouses.

6.2.8 How Does a Fuzzy Data Warehouse Behave in Comparison with a Crisp Data Warehouse?

The physical distinction between fuzzy concepts and crisp data allows the fuzzy data warehouse to be both a crisp and fuzzy data warehouse at the same time. It is possible to use the presented fuzzy data warehouse as a crisp data warehouse simply by not defining any fuzzy concepts. Even when fuzzy concepts are defined, the fuzzy data warehouse can be queried crisp. Due to the separation of fuzzy concepts, the query performance and the overall behavior of the fuzzy data warehouse corresponds exactly to a crisp data warehouse. As soon as fuzzy concepts are included in the query process, additional joins have to be done in the SQL statements. Therefore, the fuzzy concept directly impacts the query performance of the fuzzy data warehouse. However, if the fuzzy concept would have been implemented in the crisp data warehouse structure, as proposed by other approaches, it would directly impact the navigation logic or the data structure of the crisp part of the data warehouse. It can be stated that the fuzzy data warehouse behaves analogous to a crisp data warehouse for crisp data analysis and, as soon as fuzzy concepts are included in the query, it adds additional information, but also reduces query performance.

In conclusion, it can be shown that the fuzzy data warehouse improves analysis in comparison to classical crisp data warehouses. A major advantage of the fuzzy data warehouse concept in this thesis is the ability to directly integrate into existing data warehouses. The fuzzy concepts do not influence the structure of the classical data warehouse and allow querying of the data both fuzzy and crisp at the same time. This is a novelty compared to existing fuzzy data warehouse applications. Finally, with the prototype, a proof of concept is implemented that demonstrates how a fuzzy data warehouse can be realized using common data warehouse technologies. The prototype further demonstrate the ease of navigating the fuzzy data warehouse and the administration of the fuzzy concepts.

Appendices

Appendix A
The XML Schema of the Crisp Part

```
<?xml version="1.0" encoding="UTF-8"?>
<xs:schema xmlns:xs="http://www.w3.org/2001/XMLSchema"
     xmlns:fdwh="http://diuf.unifr.ch/is/fdwh"
   targetNamespace="http://diuf.unifr.ch/is/fdwh"
   elementFormDefault="qualified" attributeFormDefault
   ="qualified">
  <xs:complexType name="complexRelationType">
    <xs:sequence>
      <xs:choice>
        <xs:element name="join" type="fdwh:joinType"
            maxOccurs="1"/>
        <xs:element name="table" type="xs:string"
            maxOccurs="1"/>
        <xs:element name="relation" type="
            fdwh:complexRelationType" maxOccurs="1"/>
      </xs:choice>
      <xs:element name="key" type="xs:string" minOccurs
          ="1" maxOccurs="1"/>
    </xs:sequence>
  </xs:complexType>
  <xs:complexType name="columnType">
    <xs:sequence>
      <xs:element name="column" type="xs:string"
          minOccurs="1" maxOccurs="1"/>
      <xs:element name="display" type="xs:string"
          minOccurs="0" maxOccurs="1"/>
    </xs:sequence>
  </xs:complexType>
  <xs:complexType name="simpleRelationType">
    <xs:sequence>
```

D. Fasel, *Fuzzy Data Warehousing for Performance Measurement*, Fuzzy Management Methods, DOI 10.1007/978-3-319-04226-8,

```
      <xs:element name="table" type="xs:string"
          minOccurs="1" maxOccurs="1"/>
      <xs:element name="key" type="xs:string" minOccurs
          ="1" maxOccurs="1"/>
      <xs:element name="column" type="fdwh:columnType"
          minOccurs="0" maxOccurs="unbounded"/>
    </xs:sequence>
  </xs:complexType>
  <xs:simpleType name="aggregatorType">
    <xs:restriction base="xs:string">
      <xs:enumeration value="SUM"/>
      <xs:enumeration value="MAX"/>
      <xs:enumeration value="MIN"/>
      <xs:enumeration value="AVG"/>
      <xs:enumeration value="PROD"/>
    </xs:restriction>
  </xs:simpleType>
  <xs:simpleType name="factorType">
    <xs:restriction base="xs:string">
      <xs:enumeration value="TIMES"/>
      <xs:enumeration value="PROD"/>
    </xs:restriction>
  </xs:simpleType>
  <xs:complexType name="measureType">
    <xs:sequence>
      <xs:element name="column" type="fdwh:columnType"
          minOccurs="1" maxOccurs="1"/>
      <xs:element name="aggregator" type="
          fdwh:aggregatorType" minOccurs="1" maxOccurs="
          1"/>
    </xs:sequence>
  </xs:complexType>
  <xs:complexType name="joinType">
    <xs:sequence>
      <xs:element name="relation" type="
          fdwh:complexRelationType" minOccurs="2"
          maxOccurs="2"/>
    </xs:sequence>
  </xs:complexType>
  <xs:complexType name="bridgeType">
    <xs:sequence>
      <xs:element name="relation" type="
          fdwh:simpleRelationType" minOccurs="1"
          maxOccurs="1"/>
```

```
      <xs:element name="weight" type="xs:string"
          minOccurs="1" maxOccurs="1"/>
      <xs:element name="factor" type="fdwh:factorType"
          minOccurs="1" maxOccurs="1"/>
    </xs:sequence>
  </xs:complexType>
  <xs:complexType name="levelType">
    <xs:sequence>
      <xs:element name="name" type="xs:string"
          minOccurs="1" maxOccurs="1"/>
      <xs:element name="relation" type="
          fdwh:simpleRelationType" minOccurs="1"
          maxOccurs="1"/>
      <xs:element name="bridge" type="fdwh:bridgeType"
          minOccurs="0" maxOccurs="1"/>
    </xs:sequence>
  </xs:complexType>
  <xs:complexType name="hierarchyType">
    <xs:sequence>
      <xs:element name="name" type="xs:string"
          minOccurs="1" maxOccurs="1"/>
      <xs:element name="relation" type="
          fdwh:complexRelationType" minOccurs="1"
          maxOccurs="1"/>
      <xs:element name="level" type="fdwh:levelType"
          minOccurs="1" maxOccurs="unbounded"/>
    </xs:sequence>
  </xs:complexType>
  <xs:complexType name="dimensionType">
    <xs:sequence>
      <xs:element name="name" type="xs:string"/>
      <xs:element name="hierarchy" type="
          fdwh:hierarchyType" minOccurs="1" maxOccurs="
          unbounded"/>
    </xs:sequence>
  </xs:complexType>
  <xs:complexType name="factType">
    <xs:sequence>
      <xs:element name="relation" minOccurs="1"
          maxOccurs="1" type="fdwh:simpleRelationType"/>
      <xs:element name="measure" minOccurs="1"
          maxOccurs="unbounded" type="fdwh:measureType"/
          >
    </xs:sequence>
```

```
  </xs:complexType>
  <xs:complexType name="dimensionsType">
    <xs:sequence>
      <xs:element name="dimension" minOccurs="2"
          maxOccurs="unbounded" type="fdwh:dimensionType
          "/>
    </xs:sequence>
  </xs:complexType>
  <xs:complexType name="cubeType">
    <xs:sequence>
      <xs:element name="fact" type="fdwh:factType"
          minOccurs="1"/>
      <xs:element name="dimensions" type="
          fdwh:dimensionsType" minOccurs="1"/>
    </xs:sequence>
  </xs:complexType>
  <xs:complexType name="dwhType">
    <xs:sequence>
      <xs:element name="cube" type="fdwh:cubeType"
          minOccurs="1" maxOccurs="unbounded"/>
    </xs:sequence>
  </xs:complexType>
  <xs:element name="dwh" type="fdwh:dwhType"/>
</xs:schema>
```

Listing A.1 XML schema of crisp part

Appendix B
The XML Schema of the Fuzzy Part

```
<?xml version="1.0" encoding="UTF-8"?>
<xs:schema xmlns:fc="http://diuf.unifr.ch/is/fc"
    xmlns:xs="http://www.w3.org/2001/XMLSchema"
    attributeFormDefault="qualified" elementFormDefault
    ="qualified" targetNamespace="http://diuf.unifr.ch/
    is/fc">
  <xs:complexType name="targetType">
    <xs:choice>
      <xs:element name="column" maxOccurs="1"
          minOccurs="1"/>
      <xs:element name="query" minOccurs="1" maxOccurs
          ="1"/>
    </xs:choice>
  </xs:complexType>
  <xs:complexType name="taType">
    <xs:sequence>
      <xs:element maxOccurs="1" minOccurs="1" name="
          table" type="xs:string"/>
      <xs:element maxOccurs="1" minOccurs="1" name="
          key" type="xs:string"/>
      <xs:element maxOccurs="1" minOccurs="1" name="
          target" type="fc:targetType"/>
    </xs:sequence>
  </xs:complexType>
  <xs:complexType name="fmtType">
    <xs:sequence>
      <xs:element maxOccurs="1" minOccurs="1" name="
          table" type="xs:string"/>
    </xs:sequence>
  </xs:complexType>
```

D. Fasel, *Fuzzy Data Warehousing for Performance Measurement*, Fuzzy Management Methods, DOI 10.1007/978-3-319-04226-8,

```
<xs:complexType name="triggerType">
  <xs:sequence>
    <xs:element name="table" type="xs:string"
        maxOccurs="1" minOccurs="1"/>
    <xs:element name="name" type="xs:string"
        minOccurs="1" maxOccurs="1"/>
  </xs:sequence>
</xs:complexType>
<xs:complexType name="relationType">
  <xs:sequence>
    <xs:element maxOccurs="1" minOccurs="1" name="ta
        " type="fc:taType"/>
    <xs:element maxOccurs="1" minOccurs="1" name="
        fmt" type="xs:string"/>
    <xs:element maxOccurs="1" minOccurs="1" name="
        fct" type="xs:string"/>
    <xs:element maxOccurs="1" minOccurs="1" name="
        memfunc" type="xs:string"/>
    <xs:element maxOccurs="1" minOccurs="1" name="
        trigger" type="fc:triggerType"/>
  </xs:sequence>
</xs:complexType>
<xs:simpleType name="aggregatorType">
  <xs:restriction base="xs:string">
    <xs:enumeration value="SUM"/>
    <xs:enumeration value="MAX"/>
    <xs:enumeration value="MIN"/>
    <xs:enumeration value="AVG"/>
    <xs:enumeration value="PROD"/>
  </xs:restriction>
</xs:simpleType>
<xs:complexType name="propagationType">
  <xs:sequence>
    <xs:element maxOccurs="unbounded" minOccurs="1"
        name="fconcept" type="fc:fconceptType"/>
  </xs:sequence>
</xs:complexType>
<xs:complexType name="volatileType">
  <xs:sequence>
    <xs:element name="ta" minOccurs="1" maxOccurs="1
        " type="xs:string"/>
    <xs:element name="proc" minOccurs="1" maxOccurs=
        "1" type="xs:string"/>
  </xs:sequence>
```

```
  </xs:complexType>
  <xs:complexType name="aggregationType">
    <xs:sequence>
      <xs:element maxOccurs="1" minOccurs="0" name="
          aggregator" type="fc:aggregatorType"/>
      <xs:element maxOccurs="1" minOccurs="0" name="
          propagation" type="fc:propagationType"/>
      <xs:element maxOccurs="1" minOccurs="0" name="
          volatile" type="fc:volatileType"/>
    </xs:sequence>
  </xs:complexType>
  <xs:complexType name="termType">
    <xs:sequence>
      <xs:element maxOccurs="1" minOccurs="1" name="
          name" type="xs:string"/>
      <xs:element maxOccurs="1" minOccurs="1" name="
          key" type="xs:string"/>
    </xs:sequence>
  </xs:complexType>
  <xs:complexType name="fconceptType">
    <xs:sequence>
      <xs:element maxOccurs="1" minOccurs="1" name="
          name" type="xs:string"/>
      <xs:element maxOccurs="1" minOccurs="1" name="
          relation" type="fc:relationType"/>
      <xs:element maxOccurs="1" minOccurs="0" name="
          aggregation" type="fc:aggregationType"/>
    </xs:sequence>
  </xs:complexType>
  <xs:complexType name="conceptsType">
    <xs:sequence>
      <xs:element maxOccurs="unbounded" minOccurs="0"
          name="fconcept" type="fc:fconceptType"/>
    </xs:sequence>
  </xs:complexType>
  <xs:element name="concepts" type="fc:conceptsType"/>
</xs:schema>
```

Listing B.1 The XML schema of the fuzzy part

Appendix C
The XML Document of the Crisp Fuzzy Data Warehouse Elements


```
<?xml version="1.0" encoding="UTF-8"?>
<dwh xmlns="http://diuf.unifr.ch/is/fdwh" xmlns:xsi="
    http://www.w3.org/2001/XMLSchema-instance"
    xsi:schemaLocation="http://diuf.unifr.ch/is/fdwh␣
    file:/Users/faseldan/Documents/Uni/IS%20Group/0-
    Doktorat/xml/cube.xsd">
  <cube>
    <fact>
      <relation>
        <table>fact</table>
        <key>f_id</key>
      </relation>
      <measure>
        <column>
          <column>revenue</column>
        </column>
        <aggregator>SUM</aggregator>
      </measure>
      <measure>
        <column>
          <column>user_rating</column>
          <display>User Rating</display>
        </column>
        <aggregator>AVG</aggregator>
      </measure>
    </fact>
    <dimensions>
      <dimension>
        <name>Time</name>
        <hierarchy>
```


D. Fasel, *Fuzzy Data Warehousing for Performance Measurement*, Fuzzy Management Methods, DOI 10.1007/978-3-319-04226-8,

```
<name>Time Month</name>
<relation>
  <relation>
    <join>
      <relation>
        <table>day</table>
        <key>fk_m</key>
      </relation>
      <relation>
        <join>
          <relation>
            <table>month</table>
            <key>fk_y</key>
          </relation>
          <relation>
            <table>year</table>
            <key>y_id</key>
          </relation>
        </join>
        <key>m_id</key>
      </relation>
    </join>
    <key>d_id</key>
  </relation>
  <key>fk_d</key>
</relation>
<level>
  <name>day</name>
  <relation>
    <table>day</table>
    <key>d_id</key>
    <column>
      <column>weekday</column>
      <display>Week Day</display>
    </column>
  </relation>
</level>
<level>
  <name>month</name>
  <relation>
    <table>month</table>
    <key>m_id</key>
    <column>
      <column>monthname</column>
```

```
            <display>Month</display>
          </column>
        </relation>
      </level>
      <level>
        <name>year</name>
        <relation>
          <table>year</table>
          <key>y_id</key>
          <column>
            <column>year</column>
            <display>Year</display>
          </column>
        </relation>
      </level>
    </hierarchy>
    <hierarchy>
      <name>Time Week</name>
      <relation>
        <relation>
          <join>
            <relation>
              <table>day</table>
              <key>fk_w</key>
            </relation>
            <relation>
              <join>
                <relation>
                  <table>week</table>
                  <key>fk_y</key>
                </relation>
                <relation>
                  <table>year</table>
                  <key>y_id</key>
                </relation>
              </join>
              <key>w_id</key>
            </relation>
          </join>
          <key>d_id</key>
        </relation>
        <key>fk_d</key>
      </relation>
      <level>
```

```
          <name>day</name>
          <relation>
            <table>day</table>
            <key>d_id</key>
            <column>
              <column>weekday</column>
              <display>Week Day</display>
            </column>
          </relation>
        </level>
        <level>
          <name>week</name>
          <relation>
            <table>week</table>
            <key>w_id</key>
            <column>
              <column>week</column>
              <display>Week Number</display>
            </column>
          </relation>
        </level>
        <level>
          <name>year</name>
          <relation>
            <table>year</table>
            <key>y_id</key>
            <column>
              <column>year</column>
              <display>Year</display>
            </column>
          </relation>
        </level>
      </hierarchy>
    </dimension>
    <dimension>
      <name>Movie</name>
      <hierarchy>
        <name>Movie Category</name>
        <relation>
          <relation>
            <join>
              <relation>
                <table>movie</table>
                <key>m_id</key>
```

```
          </relation>
          <relation>
            <join>
              <relation>
                <table>mov_cat_bridge</table>
                <key>c_id</key>
              </relation>
              <relation>
                <table>category</table>
                <key>c_id</key>
              </relation>
            </join>
            <key>m_id</key>
          </relation>
        </join>
        <key>m_id</key>
      </relation>
      <key>fk_m</key>
    </relation>
    <level>
      <name>movie</name>
      <relation>
        <table>movie</table>
        <key>m_id</key>
        <column>
          <column>name</column>
          <display>Name</display>
        </column>
        <column>
          <column>overview</column>
          <display>Overview</display>
        </column>
        <column>
          <column>director</column>
          <display>Director</display>
        </column>
        <column>
          <column>actors</column>
          <display>Actors</display>
        </column>
        <column>
          <column>release_date</column>
          <display>Released</display>
        </column>
```

```
        <column>
          <column>runtime</column>
          <display>Runtime</display>
        </column>
      </relation>
    </level>
    <level>
      <name>category</name>
      <relation>
        <table>category</table>
        <key>c_id</key>
        <column>
          <column>category</column>
          <display>Category</display>
        </column>
      </relation>
      <bridge>
        <relation>
          <table>mov_cat_bridge</table>
          <key>c_id</key>
        </relation>
        <weight>weight</weight>
        <factor>PROD</factor>
      </bridge>
    </level>
  </hierarchy>
  <hierarchy>
    <name>Movie Producer</name>
    <relation>
      <relation>
        <join>
          <relation>
            <table>movie</table>
            <key>m_id</key>
          </relation>
          <relation>
            <join>
              <relation>
                <table>mov_prod_bridge</table>
                <key>p_id</key>
              </relation>
              <relation>
                <table>producer</table>
                <key>p_id</key>
```

```
          </relation>
        </join>
        <key>m_id</key>
      </relation>
    </join>
    <key>m_id</key>
  </relation>
  <key>fk_m</key>
</relation>
<level>
  <name>movie</name>
  <relation>
    <table>movie</table>
    <key>m_id</key>
    <column>
      <column>name</column>
      <display>Name</display>
    </column>
    <column>
      <column>overview</column>
      <display>Overview</display>
    </column>
    <column>
      <column>director</column>
      <display>Director</display>
    </column>
    <column>
      <column>actors</column>
      <display>Actors</display>
    </column>
    <column>
      <column>release_date</column>
      <display>Released</display>
    </column>
    <column>
      <column>runtime</column>
      <display>Runtime</display>
    </column>
  </relation>
</level>
<level>
  <name>Producer</name>
  <relation>
    <table>producer</table>
```

```
          <key>p_id</key>
          <column>
            <column>studio</column>
            <display>Producer Studio</display>
          </column>
        </relation>
        <bridge>
          <relation>
            <table>mov_prod_bridge</table>
            <key>p_id</key>
          </relation>
          <weight>weight</weight>
          <factor>PROD</factor>
        </bridge>
      </level>
    </hierarchy>
  </dimension>
  <dimension>
    <name>Type</name>
    <hierarchy>
      <name>Type</name>
      <relation>
        <relation>
          <table>type</table>
          <key>t_id</key>
        </relation>
        <key>fk_t</key>
      </relation>
      <level>
        <name>Type</name>
        <relation>
          <table>type</table>
          <key>t_id</key>
          <column>
            <column>type</column>
            <display>Type</display>
          </column>
        </relation>
      </level>
    </hierarchy>
  </dimension>
  <dimension>
    <name>Customer</name>
    <hierarchy>
```

```
        <name>Customer</name>
        <relation>
          <relation>
            <table>customer</table>
            <key>c_id</key>
          </relation>
          <key>fk_c</key>
        </relation>
        <level>
          <name>customer</name>
          <relation>
            <table>customer</table>
            <key>c_id</key>
            <column>
              <column>name</column>
              <display>Name</display>
            </column>
            <column>
              <column>bday</column>
              <display>Birth Day</display>
            </column>
            <column>
              <column>address</column>
              <display>Address</display>
            </column>
          </relation>
        </level>
      </hierarchy>
    </dimension>
    <dimension>
      <name>Employee</name>
      <hierarchy>
        <name>Employee</name>
        <relation>
          <relation>
            <table>employee</table>
            <key>e_id</key>
          </relation>
          <key>fk_e</key>
        </relation>
        <level>
          <name>Employee</name>
          <relation>
            <table>employee</table>
```

```
        <key>e_id</key>
        <column>
          <column>name</column>
          <display>Name</display>
        </column>
        <column>
          <column>bday</column>
          <display>Birth Day</display>
        </column>
      </relation>
    </level>
  </hierarchy>
</dimension>
<dimension>
  <name>Store</name>
  <hierarchy>
    <name>store</name>
    <relation>
      <relation>
        <join>
          <relation>
            <table>store</table>
            <key>fk_c</key>
          </relation>
          <relation>
            <join>
              <relation>
                <table>city</table>
                <key>fk_r</key>
              </relation>
              <relation>
                <table>region</table>
                <key>r_id</key>
              </relation>
            </join>
            <key>c_id</key>
          </relation>
        </join>
        <key>s_id</key>
      </relation>
      <key>fk_s</key>
    </relation>
    <level>
      <name>Store</name>
```

```
              <relation>
                <table>store</table>
                <key>s_id</key>
                <column>
                  <column>name</column>
                  <display>Name</display>
                </column>
                <column>
                  <column>surface</column>
                  <display>Surface</display>
                </column>
              </relation>
            </level>
            <level>
              <name>City</name>
              <relation>
                <table>city</table>
                <key>c_id</key>
                <column>
                  <column>name</column>
                  <display>Name</display>
                </column>
              </relation>
            </level>
            <level>
              <name>Region</name>
              <relation>
                <table>region</table>
                <key>r_id</key>
                <column>
                  <column>region</column>
                  <display>Region</display>
                </column>
              </relation>
            </level>
          </hierarchy>
        </dimension>
      </dimensions>
    </cube>
  </dwh>
```

Listing C.1 The XML document of the crisp fuzzy data warehouse elements

Appendix D
The XML Document of the Fuzzy Concepts


```xml
<?xml version="1.0" encoding="UTF-8"?>
<concepts xmlns="http://diuf.unifr.ch/is/fc" xmlns:xsi
    ="http://www.w3.org/2001/XMLSchema-instance"
    xsi:schemaLocation="http://diuf.unifr.ch/is/fc␣
    file:/Users/faseldan/Documents/Uni/IS%20Group/0-
    Doktorat/xml/fuzzy.xsd">
  <fconcept>
    <name>Store Surface</name>
    <relation>
      <ta>
        <table>store</table>
        <key>s_id</key>
        <target>
          <column>surface</column>
        </target>
      </ta>
      <fmt>fmt_storesurf</fmt>
      <fct>fct_storesurf</fct>
      <memfunc>fc_storesurf()</memfunc>
      <trigger>
        <table>store</table>
        <name>fc_storesurf_trigger</name>
      </trigger>
    </relation>
    <aggregation>
      <propagation>
        <fconcept>
          <name>City Store Surface</name>
          <relation>
            <ta>
```


DOI 10.1007/978-3-319-04226-8,

```
          <table>city</table>
          <key>c_id</key>
          <target>
            <query>select c_id as id, sum(surface)
                as ta from fdwh.store join fdwh.
               city on store.fk_c = city.c_id
               group by c_id</query>
          </target>
        </ta>
        <fmt>fmt_storesurf_city</fmt>
        <fct>fct_storesurf</fct>
        <memfunc>fc_storesurf_prop('city')</
           memfunc>
        <trigger>
          <table>store</table>
          <name>fc_storesurf_city_trigger</name>
        </trigger>
      </relation>
    </fconcept>
    <fconcept>
      <name>Region store surface</name>
      <relation>
        <ta>
          <table>region</table>
          <key>r_id</key>
          <target>
            <query>select r_id, sum(surface) as ta
                from fdwh.store join fdwh.city on
               store.fk_c = city.c_id join fdwh.
               region on city.fk_r = region.r_id
               group by r_id</query>
          </target>
        </ta>
        <fmt>fmt_storesurf_region</fmt>
        <fct>fct_storesurf</fct>
        <memfunc>fct_storesurf_prop('region')</
           memfunc>
        <trigger>
          <table>store</table>
          <name>fc_storesurf_region_trigger</name>
        </trigger>
      </relation>
    </fconcept>
  </propagation>
```

```
    </aggregation>
  </fconcept>
  <fconcept>
    <name>Customer Revenue</name>
    <relation>
      <ta>
        <table>customer</table>
        <key>c_id</key>
        <target>
          <query>select c_id as id, sum(revenue) as ta
               from fdwh.fact join fdwh.customer on
              fact.fk_c = customer.c_id group by c_id</
              query>
        </target>
      </ta>
      <fmt>fmt_custrev</fmt>
      <fct>fct_custrev</fct>
      <memfunc>fc_custrev()</memfunc>
      <trigger>
        <table>fact</table>
        <name>fc_custrev_trigger</name>
      </trigger>
    </relation>
  </fconcept>
  <fconcept>
    <name>Customer Age</name>
    <relation>
      <ta>
        <table>customer</table>
        <key>c_id</key>
        <target>
          <query>select c_id as id, extract(years from
               age(bday)) as ta from fdwh.customer</
              query>
        </target>
      </ta>
      <fmt>fmt_custage</fmt>
      <fct>fct_custage</fct>
      <memfunc>fc_custage()</memfunc>
      <trigger>
        <table>customer</table>
        <name>fc_custage_trigger</name>
      </trigger>
    </relation>
```

```
</fconcept>
<fconcept>
  <name>Employee Age</name>
  <relation>
    <ta>
      <table>employee</table>
      <key>e_id</key>
      <target>
        <query>select e_id as id, extract(years from
             age(bday)) as ta from fdwh.employee</
            query>
      </target>
    </ta>
    <fmt>fmt_emplage</fmt>
    <fct>fct_emplage</fct>
    <memfunc>fc_emplage()</memfunc>
    <trigger>
      <table>employee</table>
      <name>fc_emplage_trigger</name>
    </trigger>
  </relation>
</fconcept>
<fconcept>
  <name>User Rating</name>
  <relation>
    <ta>
      <table>fact</table>
      <key>f_id</key>
      <target>
        <column>user_rating</column>
      </target>
    </ta>
    <fmt>fmt_rating</fmt>
    <fct>fct_rating</fct>
    <memfunc>fc_rating()</memfunc>
    <trigger>
      <table>fact</table>
      <name>fc_rating_trigger</name>
    </trigger>
  </relation>
  <aggregation>
    <aggregator>AVG</aggregator>
  </aggregation>
</fconcept>
```

```
<fconcept>
  <name>Movie Genre</name>
  <relation>
    <ta>
      <table>movie</table>
      <key>m_id</key>
      <target>
        <column>m_id</column>
      </target>
    </ta>
    <fmt>fmt_movcat</fmt>
    <fct>fct_movcat</fct>
    <memfunc/>
    <trigger>
      <table/>
      <name/>
    </trigger>
  </relation>
  <aggregation>
    <aggregator>AVG</aggregator>
  </aggregation>
</fconcept>
<fconcept>
  <name>Revenue</name>
  <relation>
    <ta>
      <table>fact</table>
      <key>f_id</key>
      <target>
        <column>revenue</column>
      </target>
    </ta>
    <fmt>fmt_revenue</fmt>
    <fct>fct_revenue</fct>
    <memfunc>fc_revenue</memfunc>
    <trigger>
      <table>revenue</table>
      <name>fc_revenue_trigger</name>
    </trigger>
  </relation>
  <aggregation>
    <propagation>
      <fconcept>
        <name>Movie Revenue</name>
```

```
<relation>
  <ta>
    <table>movie</table>
    <key>m_id</key>
    <target>
      <query>select m_id as id, sum(revenue)
           as ta from fdwh.fact join fdwh.
          movie on fact.fk_m = movie.m_id
          group by m_id</query>
    </target>
  </ta>
  <fmt>fmt_revenue_movie</fmt>
  <fct>fct_revenue</fct>
  <memfunc>fc_revenue_prop('movie')</memfunc
     >
  <trigger>
    <table>revenue</table>
    <name>fc_revenue_movie_trigger</name>
  </trigger>
</relation>
</fconcept>
<fconcept>
  <name>Category Revenue</name>
  <relation>
    <ta>
      <table>category</table>
      <key>c_id</key>
      <target>
        <query>select category.c_id as id, sum
            (revenue) as ta from fdwh.fact join
             fdwh.movie on fact.fk_m = movie.
            m_id join fdwh.mov_cat_bridge on
            movie.m_id = mov_cat_bridge.m_id
            join fdwh.category on
            mov_cat_bridge.c_id = category.c_id
             group by category.c_id</query>
      </target>
    </ta>
    <fmt>fmt_revenue_category</fmt>
    <fct>fct_revenue</fct>
    <memfunc>fc_revenue_prop('category')</
       memfunc>
    <trigger>
      <table>revenue</table>
```

```
        <name>fc_revenue_category_trigger</name>
      </trigger>
    </relation>
  </fconcept>
  <fconcept>
    <name>Producer Revenue</name>
    <relation>
      <ta>
        <table>producer</table>
        <key>p_id</key>
        <target>
          <query>
  SELECT producer.p_id AS id, SUM(revenue) AS
       ta FROM fdwh.fact INNER JOIN fdwh.movie
       ON fdwh.fact.fk_m=fdwh.movie.m_id
  INNER JOIN fdwh.mov_cat_bridge ON fdwh.
      movie.m_id=fdwh.mov_cat_bridge.m_id
  INNER JOIN fdwh.mov_prod_bridge ON fdwh.
      movie.m_id=fdwh.mov_prod_bridge.m_id
  INNER JOIN fdwh.producer ON fdwh.
      mov_prod_bridge.p_id=fdwh.producer.p_id
  GROUP BY producer.p_id
 </query>
        </target>
      </ta>
      <fmt>fmt_revenue_producer</fmt>
      <fct>fct_revenue</fct>
      <memfunc>fc_revenue_prop('producer')</
         memfunc>
      <trigger>
        <table>revenue</table>
        <name>fc_revenue_producer_trigger</name>
      </trigger>
    </relation>
  </fconcept>
  <fconcept>
    <name>Employee Revenue</name>
    <relation>
      <ta>
        <table>employee</table>
        <key>e_id</key>
        <target>
          <query>select e_id as id, sum(revenue)
               as ta from fdwh.fact join fdwh.
```

```
            employee on fact.fk_e = employee.
            e_id group by e_id</query>
      </target>
    </ta>
    <fmt>fmt_revenue_employee</fmt>
    <fct>fct_revenue</fct>
    <memfunc>fc_revenue_prop('employee')</
       memfunc>
    <trigger>
      <table>revenue</table>
      <name>fc_revenue_employee_trigger</name>
    </trigger>
  </relation>
</fconcept>
<fconcept>
  <name>Type Revenue</name>
  <relation>
    <ta>
      <table>type</table>
      <key>t_id</key>
      <target>
        <query>select t_id as id, sum(revenue)
             as ta from fdwh.fact join fdwh.
            type on fact.fk_t = type.t_id group
             by t_id</query>
      </target>
    </ta>
    <fmt>fmt_revenue_type</fmt>
    <fct>fct_revenue</fct>
    <memfunc>fc_revenue_prop('type')</memfunc>
    <trigger>
      <table>revenue</table>
      <name>fc_revenue_type_trigger</name>
    </trigger>
  </relation>
</fconcept>
<fconcept>
  <name>Customer Revenue Propagated</name>
  <relation>
    <ta>
      <table>customer</table>
      <key>c_id</key>
      <target>
```

```
          <query>select c_id as id, sum(revenue)
              as ta from fdwh.fact join fdwh.
              customer on fact.fk_c = customer.
              c_id group by c_id</query>
        </target>
      </ta>
      <fmt>fmt_revenue_customer</fmt>
      <fct>fct_revenue</fct>
      <memfunc>fc_revenue_prop('customer')</
         memfunc>
      <trigger>
        <table>revenue</table>
        <name>fc_revenue_customer_trigger</name>
      </trigger>
    </relation>
  </fconcept>
  <fconcept>
    <name>Store Revenue</name>
    <relation>
      <ta>
        <table>store</table>
        <key>s_id</key>
        <target>
          <query>select s_id as id, sum(revenue)
              as ta from fdwh.fact join fdwh.
              store on fact.fk_s = store.s_id
              group by s_id</query>
        </target>
      </ta>
      <fmt>fmt_revenue_store</fmt>
      <fct>fct_revenue</fct>
      <memfunc>fc_revenue_prop('store')</memfunc
         >
      <trigger>
        <table>revenue</table>
        <name>fc_revenue_store_trigger</name>
      </trigger>
    </relation>
  </fconcept>
  <fconcept>
    <name>City Revenue</name>
    <relation>
      <ta>
        <table>city</table>
```

```
        <key>c_id</key>
        <target>
          <query>select c_id as id, sum(revenue)
               as ta from fdwh.fact join fdwh.
              store on fact.fk_s = store.s_id
              join fdwh.city on store.fk_c = city
              .c_id group by c_id</query>
        </target>
      </ta>
      <fmt>fmt_revenue_city</fmt>
      <fct>fct_revenue</fct>
      <memfunc>fc_revenue_prop('city')</memfunc>
      <trigger>
        <table>revenue</table>
        <name>fc_revenue_city_trigger</name>
      </trigger>
    </relation>
  </fconcept>
  <fconcept>
    <name>Region Revenue</name>
    <relation>
      <ta>
        <table>region</table>
        <key>r_id</key>
        <target>
          <query>select r_id as id, sum(revenue)
               as ta from fdwh.fact join fdwh.
              store on fact.fk_s = store.s_id
              join fdwh.city on store.fk_c = city
              .c_id join fdwh.region on city.fk_r
               = region.r_id group by r_id</query
              >
        </target>
      </ta>
      <fmt>fmt_revenue_region</fmt>
      <fct>fct_revenue</fct>
      <memfunc>fc_revenue_prop('region')</
         memfunc>
      <trigger>
        <table>revenue</table>
        <name>fc_revenue_region_trigger</name>
      </trigger>
    </relation>
  </fconcept>
```

```
<fconcept>
  <name>Day Revenue</name>
  <relation>
    <ta>
      <table>day</table>
      <key>s_id</key>
      <target>
        <query>select d_id as id, sum(revenue)
             as ta from fdwh.fact join fdwh.day
             on fact.fk_d = day.d_id group by
            d_id</query>
      </target>
    </ta>
    <fmt>fmt_revenue_day</fmt>
    <fct>fct_revenue</fct>
    <memfunc>fc_revenue_prop('day')</memfunc>
    <trigger>
      <table>revenue</table>
      <name>fc_revenue_day_trigger</name>
    </trigger>
  </relation>
</fconcept>
<fconcept>
  <name>Month Revenue</name>
  <relation>
    <ta>
      <table>month</table>
      <key>m_id</key>
      <target>
        <query>select m_id as id, sum(revenue)
             as ta from fdwh.fact join fdwh.day
             on fact.fk_d = day.d_id join fdwh.
            month on day.fk_m = month.m_id
            group by m_id</query>
      </target>
    </ta>
    <fmt>fmt_revenue_month</fmt>
    <fct>fct_revenue</fct>
    <memfunc>fc_revenue_prop('month')</memfunc
       >
    <trigger>
      <table>revenue</table>
      <name>fc_revenue_month_trigger</name>
    </trigger>
```

```
    </relation>
  </fconcept>
  <fconcept>
    <name>Week Revenue</name>
    <relation>
      <ta>
        <table>week</table>
        <key>w_id</key>
        <target>
          <query>select w_id as id, sum(revenue)
               as ta from fdwh.fact join fdwh.day
               on fact.fk_d = day.d_id join fdwh.
             week on day.fk_w = week.w_id group
             by w_id</query>
        </target>
      </ta>
      <fmt>fmt_revenue_week</fmt>
      <fct>fct_revenue</fct>
      <memfunc>fc_revenue_prop('week')</memfunc>
      <trigger>
        <table>revenue</table>
        <name>fc_revenue_week_trigger</name>
      </trigger>
    </relation>
  </fconcept>
  <fconcept>
    <name>Year Revenue</name>
    <relation>
      <ta>
        <table>year</table>
        <key>y_id</key>
        <target>
          <query>select y_id as id, sum(revenue)
               as ta from fdwh.fact join fdwh.day
               on fact.fk_m = day.d_id join fdwh.
             month on day.fk_m = month.m_id join
               fdwh.year on month.fk_y = year.
             y_id group by y_id</query>
        </target>
      </ta>
      <fmt>fmt_revenue_year</fmt>
      <fct>fct_revenue</fct>
      <memfunc>fc_revenue_prop('year')</memfunc>
      <trigger>
```

```
                <table>revenue</table>
                <name>fc_revenue_year_trigger</name>
              </trigger>
            </relation>
          </fconcept>
        </propagation>
        <volatile>
          <ta>revenue</ta>
          <proc>fc_revenue_volatile</proc>
        </volatile>
      </aggregation>
    </fconcept>
  </concepts>
```

Listing D.1 The XML document of the fuzzy concepts

Bibliography

[AAG+96] S. Agrawal, R. Agrawal, E.A. Gupta, J.F. Naughton, R. Ramakrishnan, S. Sarawagi, On the computation of multidimensional aggregates, in *Proceedings of the 22th International Conference on Very Large Data Bases*, Hong Kong (Morgan Kaufmann, 1996), pp. 506–521

[AGS95] R. Agrawal, A. Gupta, S. Sarawagi, Modeling multidimensional databases, research report. Technical report, IBM Almaden Research Center, San Jose, 1995

[AGS97] R. Agrawal, A. Gupta, S. Sarawagi, Modeling multidimensional databases, in *Proceedings of the Thirteenth International Conference on Data Engineering*, Birmingham (IEEE Computer Society, 1997), pp. 232–243

[AK03] R. Alhajj, M. Kaya, Integrating fuzziness into OLAP for multidimensional fuzzy association rules mining, in *Proceedings of the Third IEEE International Conference on Data Mining*, Melbourne, 2003

[Als85] C. Alsina, On a family of connectives for fuzzy sets. Fuzzy Sets Syst. **16**(3), 231–235 (1985)

[Ass11] Web Analytics Association, http://www.webanalyticsassociation.org, Nov 2011

[BDJ+05] D. Burdick, P.M. Deshpande, T.S. Jayram, R. Ramakrishnan, S. Vaithyanathan, OLAP over uncertain and imprecise data, in *Proceedings of the 31st International Conference on Very Large Data Bases*, Trondheim, 2005, pp. 970–981

[CCS93] E.F. Codd, S.B. Codd, C.T. Smalley, *Providing OLAP to User-Analysts: An IT Mandate*, Technical report, Codd & Associates, 1993

[CD97] S. Chaudhuri, U. Dayal, An overview of data warehousing and OLAP technology. ACM SIGMOD Rec. **26**(1), 65–74 (1997)

[Che76] P.P. Chen, The entity-relationship model towards unified view of data. ACM Trans. Database Syst. (TODS) **1**, 9–36 (1976)

[Cod70] E.F. Codd, A relational model for large shared data banks. Commun. ACM **13**(6), 377–387 (1970)

[Cod71] E.F. Codd, Further normalization of the database relational model, in *Database Systems Courant Computer Science Symposium*, vol. 6, ed. by R. Rustin (Prentice-Hall, Englewood Cliffs, 1971)

[Cod87] E.F. Codd, More commentary on missing information in relational databases (applicable and inapplicable information). ACM SIGMOD Rec. **16**, 42–50 (1987)

[CS94] S. Chaudhuri, K. Shim, Including group-by in query optimization, in *Proceedings of the International Conference on Very Large Data Bases*, Santiago (IEEE, 1994), pp. 354–366

[CS96] S. Chaudhuri, K. Shim, Optimizing queries with aggregate views, in *Proceedings of the 5th International Conference on Extending Database Technology: Advances in Database Technology*, Avignon (Springer, 1996), pp. 167–182

D. Fasel, *Fuzzy Data Warehousing for Performance Measurement*, Fuzzy Management Methods, DOI 10.1007/978-3-319-04226-8,

[CS99] P. Chamoni, S. Stock, Temporal structures in data warehousing, in *Data Warehousing and Knowledge Discovery*. Lecture Notes in Computer Science, vol. 1676, ed. by M. Mohania, A.M. Tjoa (Springer, Berlin/Heidelberg, 1999), pp. 353–358

[DM88] B.A. Devlin, P.T. Murphy, An architecture for a business and information system. IBM Syst. J. **27**(1), 60–80 (1988)

[DMS+04] M. Delgado, C. Molina, D. Sanchez, A. Vila, L. Rodriguez-Ariza, A fuzzy multidimensional model for supporting imprecision in OLAP, in *IEEE International Conference on Fuzzy Systems*, Budapest, vol. 3 (IEEE, 2004), pp. 1331–1336

[DP80] D. Dubois, H.M. Prade, *Fuzzy Sets and Systems: Theory and Applications*, vol. 144 (Academic, New York, 1980)

[DP85] D. Dubois, H.M. Prade, A review of fuzzy set aggregation connectives. Inf. Sci. **36**(1–2), 85–121 (1985)

[Eck95] W.W. Eckerson, Three tier client/server architectures: achieving scalability, performance, and efficiency in client/server applications. Open Inf. Syst. **10**(1), 3–20 (1995)

[Ede01] J. Eder, C. Koncilia, Changes of dimension data in temporal data warehouses, in *Data Warehousing and Knowledge Discovery* (Springer, Heidelberg, 2001), pp. 284–293

[Fas09] D. Fasel, A fuzzy data warehouse approach for the customer performance measurement for a hearing instrument manufacturing company, in *Proceedings of Fuzzy Systems and Knowledge Discovery*, Tianjin (IEEE Explore, 2009)

[FD03] L. Feng, T.S. Dillon, Using fuzzy linguistic representations to provide explanatory semantics for data warehouses. IEEE Trans. Knowl. Data Eng. **15**(1), 86–102 (2003)

[FS10] D. Fasel, K. Shahzad, A data warehouse model for integrating fuzzy concepts in meta table structures, in *17th IEEE International Conference and Workshops on the Engineering of Computer-Based Systems*, Oxford (IEEE Computer Society, 2010), pp. 100–109

[FS12] D. Fasel, K. Shahzad, Fuzzy data warehouse for performance analysis, in *Fuzzy Methods for Customer Relationship Management and Marketing: Applications and Classifications*, ed. by A. Meier, L. Donzé (IGP Global, Hershey, 2012), pp. 217–251

[FZ09] D. Fasel, D. Zumstein, A fuzzy data warehouse approach for web analytics, in *Visioning and Engineering the Knowledge Society – A Web Science Perspective*, ed. by M.D. Lytras, E. Damiani, J.M. Carroll, R.D. Tennyson, D. Avison, A. Naeve, A. Dale, P. Lefrere, F. Tan, J. Sipior, G. Vossen. Volume 5736 of Lecture Notes in Computer Science (Springer, Berlin, 2009), pp. 276–285

[GbGA06] J. Gethland, B. Galbraith, D. Almaer, *Pragmatic Ajax – A Web 2.0 Primer* (The Pragmatic Bookshelf, Raleigh, 2006)

[GCB+97] J. Gray, S. Chaudhuri, A. Bosworth, A. Layman, D. Reichart, M. Venkatrao, F. Pellow, H. Pirahesh, Data cube: a relational aggregation operator generalizing group-by, cross-tab, and sub-totals. Data Min. Knowl. Discov. **1**(1), 29–53 (1997)

[Gil76] R. Giles, Lukasiewicz logic and fuzzy set theory+. Int. J. Man-Machine Stud. **8**(3), 313–327 (1976)

[Gon09] A. Goncalves, *Beginning Java EE 6 Platform with GlassFish 3* (Apress, New York, 2009)

[GUP04] J. Galindo, A. Urrutia, M. Piattini, Representation of fuzzy knowledge in relational databases, in *Proceedings of the 15th International Workshop on Database and Expert Systems Applications*, Zaragoza (IEEE, 2004), pp. 917–921

[Ham78] H. Hamacher, *Über logische Aggregationen nicht-binär explizierter Entscheidungskriterien* (Fischer Verlag, Frankfurt, 1978)

[Has08] M. Hassler, *Web Analytics* (MITP, Heidelberg, 2008)

[Hef07] D.R. Heffelfinger, *Java EE Development Using GlassFish Application Server* (Packt Publishing, Birmingham, 2007)

[Hef11] D.R. Heffelfinger, *Java EE Development with NetBeans 6* (Packt Publishing, Birmingham, 2011)

[HKYC04] H.-G. Hwang, C.-Y. Ku, D.C. Yen, C.-C. Cheng, Critical factors influencing the adoption of data warehouse technology: a study of the banking industry in Taiwan. Decis. Support Syst. **37**(1), 1–21 (2004)

[HM05] E.R. Harold, W.S. Means, *XML in a Nutshell*, vol. 3 (O'Reilly, Sebastopol, 2005)

[HMPR04] A.R. Hevner, S.T. March, J. Park, S. Ram, Design science in information systems research. MIS Q. **28**(1), 75–105 (2004)

[HR04] D. Harel, B. Rumpe, Meaningful modeling: what's the semantics of "Semantics"? Computer **37**(10), 64–72 (2004)

[IMD11] IMDb.com, Inc., The internet movie database. http://www.imdb.com/, Sept 2011

[INC11a] ICEsoft Technologies Inc., ICEFaces. http://www.icefaces.org/main/home, Nov 2011

[Inc11b] Oracle America Inc., Java architecture for XML binding. http://jaxb.java.net, Nov 2011

[Inc11c] Oracle America Inc., Java server faces. http://www.oracle.com/technetwork/java/javaee/javaserverfaces-139869.html, Nov 2011

[Inc11d] VMWare Inc., JDBCTemplate documentation. http://static.springsource.org/spring/docs/2.0.x/reference/jdbc.html, Nov 2011

[Inc11e] VMWare Inc., Spring framework. http://www.springsource.org, Nov 2011

[Inm05] W.H. Inmon, *Building the Data Warehouse*, 4th edn. (Wiley, Indianapolis, 2005)

[Ion08] A. Ionas, Modellierung, Entwicklung und Nutzung eines Data Warehouse für medizinische Communication Centers. PhD thesis, Departement für Informatik – Universität Fribourg, Fribourg, 2008

[ISN08] W.H. Inmon, D. Strauss, G. Neushloss, *DW 2.0 – The Architecture for the Next Generation of Data Warehousing* (Morgan Kaufmann, Amsterdam, 2008)

[ISO08] ISO, ISO/IEC 9075(1-4,9-11,13,14):2008 – Information Technology – Database Languages – SQL, Geneva, 2008

[Kas96] N.K. Kasabov, *Foundations of Neural Networks, Fuzzy Systems, and Knowledge Engineering* (MIT, Cambridge, 1996)

[Kau07] A. Kaushik, *Web Analytics – An Hour a Day* (O'Reilly, New York, 2007)

[Kau12] M. Kaufmann, Inductive fuzzy classification – yet to publish. PhD thesis, University of Fribourg, Fribourg, 2012

[KC04] R. Kimball, J. Caserta, *The Data Warehouse ETL Toolkit* (Wiley, Indianapolis, 2004)

[KI93] B. Kosko, S. Isaka, Fuzzy logic. Sci. Am. **269**(1), 76–81 (1993)

[Kim98] R. Kimball, Help for dimensional modeling: helper tables let you design and manage multivalued dimensions successfully. *DBMS online* (Miller Freeman Inc., 1998), http://www.kimballgroup.com/1998/08/02/help-for-dimensional-modeling/

[KK07] D.P.V. Kasinadh, P. Radha Krishna, Building fuzzy OLAP using multi-attribute summarization, in *Proceedings of the International Conference on Computational Intelligence and Multimedia Applications (ICCIMA 2007)*, Sivakasi, vol. 01 (IEEE, 2007), pp. 370–374

[KKD05] K.V.N.N. Pavan Kumar, P. Radha Krishna, S. Kumar De, Fuzzy OLAP cube for qualitative analysis, in *Intelligent Sensing and Information Processing*, Chennai, 2005, pp. 290–295

[Kos90] B. Kosko, Fuzziness vs. probability. Int. J. Gen. Syst. **17**(2), 211–240 (1990)

[Kos94a] B. Kosko, Fuzzy systems as universal approximators. IEEE Trans. Comput. **43**(11), 1329–1333 (1994)

[Kos94b] B. Kosko, The probability monopoly. IEEE Trans. Fuzzy Syst. **2**(1), 32–33 (1994)

[KR02] R. Kimball, M. Ross, *The Data Warehouse Toolkit* (Wiley, New York, 2002)

[KRT+08] R. Kimball, M. Ross, W. Thornthwaite, J. Mundy, B. Becker, *The Data Warehouse Lifecycle Toolkit* (Wiley, Indianapolis, 2008)

[KW99] N. Kasabov, B. Woodford, Rule insertion and rule extraction from evolving fuzzy neural networks: algorithms and applications for building adaptive, intelligent expert systems, in *IEEE International Fuzzy Systems Conference Proceedings*, Seoul, vol. 3 (IEEE, 1999), pp. 1406–1411

[Leh98] W. Lehner, Modeling large scale OLAP scenarios, in *Advances in Database Technology – EDBT'98*, Valencia, vol. 1377 (Springer, 1998), pp. 153–167

[Leh03] W. Lehner, *Datenbanktechnologien für Data-Warehouse-Systeme – Konzepte und Methoden* (dpunkt.verlag, Heidelberg, 2003)

[LL03] M. Levene, G. Loizou, Why is the snowflake schema a good data warehouse design? Inf. Syst. **28**(3), 225–240 (2003)

[LS97] H.-J. Lenz, A. Shoshani, Summarizability in OLAP and statistical data bases, in *Statistical and Scientific Database Management*, Olympia, 1997

[LT01] H.J. Lenz, B. Thalheim, OLAP databases and aggregation functions, in *Proceedings of the Thirteenth International Conference on Scientific and Statistical Database Management, SSDBM 2001*, Fairfax (IEEE Computer Society, 2001), p. 0091

[Mic10] Microsoft Corporation, MDX specifications. http://msdn.microsoft.com/en-us/library/Aa216767, Apr 2010

[MMWS03] A. Meier, C. Mezger, N. Werro, G. Schindler, Zur unscharfen Klassifikation von Datenbanken mit fCQL, in *Proceedings of the GI-Workshop LLWA Lernen, Lernen, Wissen, Adaptivität*, Karlsruhe, 2003

[MRASV06] C. Molina, L. Rodríguez-Ariza, D. Sánchez, M.A. Vila, A new fuzzy multidimensional model. IEEE Trans. Fuzzy Syst. **14**(6), 897–912 (2006)

[MS95] S.T. March, G.F. Smith, Design and natural science research on information technology. Decis. Support Syst. **15**(4), 251–266 (1995)

[MS97] A. Motro, P. Smets, *Uncertainty Management in Information Systems: From Needs to Solutions* (Kluwer Academic, Boston, 1997)

[MSSV01] A. Meier, C. Savary, G. Schindler, Y. Veryha, Database schema with fuzzy classification and classification query language, in *Proceedings of the International Congress on Computational Intelligence – Methods and Applications*, University of Wales, Bangor, 2001

[MSW07] A. Meier, G. Schindler, N. Werro, Extending relational databases with fuzzy classification. Department of Informatics, University of Fribourg, Fribourg, 2007

[MSW08] A. Meier, G. Schindler, N. Werro, Fuzzy classification on relational databases, in *Handbook of Research on Fuzzy Information Processing in Databases*, vol. 2, ed. by J. Galindo (IGI Global, Hershey, 2008), pp. 586–614

[MT81] M. Mizumoto, K. Tanaka, Fuzzy sets and their operations. Inf. Control **48**(1), 30–48 (1981)

[MT02] E. Medina, J. Trujillo, A standard for representing multidimensional properties: the common warehouse model (CWM), in *Advances in Databases and Information Systems*, ed. by Y. Manolopoulos, P. Navrat. Volume 2435/2002 of Lecture Notes in Computer Science (Springer, Berlin, 2002), pp. 232–247

[MV00] A.O. Mendelzon, A.A. Vaisman, Temporal queries in OLAP, in *Proceedings of the 26th International Conference on Very Large Data Bases*, Cairo (Morgan Kaufmann, 2000), pp. 242–253

[MWAS05] A. Meier, N. Werro, M. Albrecht, M. Sarakinos, Using a fuzzy classification query language for customer relationship management, in *Proceedings of the 31st VLDB Conference*, Trondheim, 2005

[MZ04] E. Malinowski, E. Zimányi, Representing spatiality in a conceptual multidimensional model, in *Proceedings of the 12th Annual ACM International Workshop on Geographic Information Systems*, Washington, DC (ACM, 2004), pp. 12–22

[Nau00] D.D. Nauck, Data analysis with neuro-fuzzy methods. Technical report, Fakultät für Informatik der Otto-von-Guericke-Universität Magdeburg, 2000

[NH94] R. Ng, J. Han, Efficient and effective clustering method for spatial data mining, in *Proceedings of the 20th International Conference on Very Large Databases*, Santiago, 1994, pp. 144–155

[Nüe11] S. Nüesch, OLAP cube for a fuzzy data warehouse. Bachelor thesis, University of Fribourg, Fribourg, 2011

[OLA10] OLAP Council, OLAP and OLAP server definitions. http://www.olapcouncil.org/research/resrchly.htm, Aug 2010

[Ora10] Oracle, Oracle essbase hyperion. http://www.oracle.com/technetwork/middleware/essbase/overview/index.html, Jan 2010

[Ora11] Oracle, Oracle essbase hyperion database administration documentation. http://download.oracle.com/docs/cd/E17236_01/epm.1112/esb_dbag/launch.html, Sept 2011

[Pen10] Pentaho Corporation, Mondrian OLAP server. http://mondrian.pentaho.org/, Apr 2010

[Per06] B.W. Perry, *Ajax Hacks* (O'Reilly, Sebastopol, 2006)

[Pet94] S. Peterson, Stars: a pattern language for query optimized schema. Sequent Computer Systems, 1994. http://c2.com/ppr/stars.html

[PJ01] T.B. Pedersen, C.S. Jensen, Multidimensional database technology. Computer **34**(12), 40–46 (2001)

[PJD99] T.B. Pedersen, C.S. Jensen, C.E. Dyreseon, Supporting imprecision in multidimensional databases using granularities, in *Eleventh International Conference on Scientific and Statistical Database Management*, Cleveland, 1999

[Pos11a] PostgreSQL Global Development Group, PostgreSQL. http://www.postgresql.org/, Jan 2011

[Pos11b] PostgreSQL Global Development Group, PostgreSQL documentation. http://www.postgresql.org/docs/current/static, Jan 2011

[Pos11c] PostgreSQL Global Development Group, PostgreSQL documentation of server programming. http://www.postgresql.org/docs/current/static/server-programming.html, Sept 2011

[Pos11d] PostgreSQL Global Development Group. PostgreSQL documentation of window functions. http://www.postgresql.org/docs/current/static/functions-window.html, June 2011

[PSP07] D. Pérez, M.J. Somodevilla, I.H. Pineda, Fuzzy spatial data warehouse: a multidimensional model, in *Eighth Mexican International Conference on Current Trends in Computer Science*, Morelia, 2007

[Pur02] T. Purcell, Star join optimization: DB2 UDB for z/OS and OS/390. Int. DB2 User Group (IDUG) Solut. J. **9**(1), 17–19 (2002)

[Rat11] C. Rathgeb, Entwicklung eines fuzzy data warehouse. Bachelor thesis, University of Fribourg, Fribourg, 2011

[RS90] M. Rafanelli, A. Shoshani, STORM: a statistical object representation model, in *Statistical and Scientific Data Base Management Conference*, Charlotte, 1990, pp. 14–29

[Sch98] G. Schindler, Fuzzy-Datenanalyse durch kontextbasierte Datenbankabfragen. PhD thesis, Technische Hochschule Aachen, 1998

[Sch10] L. Scheuner, Entwurf und Erstellung eines Data Warehouse für die Schweizerische Futtermitteldatenbank. Master's thesis, Universität Zürich and Universität Fribourg, 2010

[Sho82] A. Shoshani, Statistical databases: characteristics, problems, and some solutions, in *Proceedings of the 8th International Conference on Very Large Data Bases*, Mexico City (Morgan Kaufmann, 1982), pp. 208–222

[SMH04] H. Schepperle, A. Merkel, A. Haag, Erhalt von imperfektion in einem data warehouse, in *Internationales Symposium: Data-Warehouse-Systeme und Knowledge-Discovery*, Zaragoza, 2004

[SSR08] L. Sapir, A. Shmilovici, L. Rokach, A methodology for the design of a fuzzy data warehouse, in *4th International IEEE Conference on Intelligent Systems, IS'08*, Varna, vol. 1, 2008

[Ter11] Teradata, http://www.teradata.com, June 2011

[Til91] T. Tilli, *Fuzzy-Logik: Grundlagen, Anwendungen, Hard- und Software* (Franzis-Verlag GmbH, München, 1991)

[TMD10] TMDb Inc., The open movie database project. http://www.themoviedb.org/, Dec 2010

[TZZ79] U. Thole, H.J. Zimmermann, P. Zysno, On the suitability of minimum and product operators for the intersection of fuzzy sets* 1. Fuzzy Sets Syst. **2**(2), 167–180 (1979)

[Vas98] P. Vassiliadis, Modeling multidimensional databases, cubes and cube operations, in *Proceedings of the 10th International Conference on Scientific and Statistical Database Management*, Capri, pp. 53–62 (IEEE, 1998)

[Ver11] Vertica, Vertica database. http://www.vertica.com/, Sept 2011

[VS99] P. Vassiliadis, T. Sellis, A survey of logical models for OLAP databases. ACM SIGMOD Rec. **28**(4), 64–69 (1999)

[VS00] P. Vassiliadis, S. Skiadopoulos, Modelling and optimisation issues for multidimensional databases, in *Advanced Information Systems Engineering* (Springer, Berlin, 2000), pp. 482–497

[Wer88] B. Werners, Aggregation models in mathematical programming. Math. Models Decis. Support **48**, 295–305 (1988)

[Wer08] N. Werro, Fuzzy classification of online customers. PhD thesis, University of Fribourg, Fribourg, 2008

[Yag80] R.R. Yager, On a general class of fuzzy connectives. Fuzzy Sets Syst. **4**(3), 235–242 (1980)

[Zad65] L.A. Zadeh, Fuzzy sets. Inf. Control **8**, 338–353 (1965)

[Zad75a] L.A. Zadeh, The concept of a linguistic variable and its application to approximate reasoning – Part I. Inf. Sci. **8**, 199–249 (1975)

[Zad75b] L.A. Zadeh, The concept of a linguistic variable and its application to approximate reasoning – Part II. Inf. Sci. **8**, 301–357 (1975)

[Zad75c] L.A. Zadeh, The concept of a linguistic variable and its application to approximate reasoning – Part III. Inf. Sci. **9**, 43–80 (1975)

[Zad78a] L.A. Zadeh, Fuzzy sets as a basis for a theory of possibility. Fuzzy Sets Syst. **1**, 3–28 (1978)

[Zad78b] L.A. Zadeh, PRUF – a meaning representation language for natural languages. Int. J. Man-Machine Stud. **10**, 395–460 (1978)

[Zad83] L.A. Zadeh, A computational approach to fuzzy quantifiers in natural languages* 1. Comput. Math. Appl. **9**(1), 149–184 (1983)

[Zad94] L.A. Zadeh, Fuzzy logic, neural networks, and soft computing. Commun. ACM **37**(3), 77–84 (1994)

[Zad96] L.A. Zadeh, Test-score semantics for natural languages and meaning representation via PRUF, in *Fuzzy Sets, Fuzzy Logic, and Fuzzy Systems: Selected Papers*, vol. 6 (World Scientific, Singapore, 1996)

[Zim91] H.-J. Zimmermann, *Fuzzy Set Theory and its Applications* (Kluwer Academic, Boston, 1991)

[ZKY96] L.A. Zadeh, G.J. Klir, B. Yuan (eds.), *Fuzzy Sets, Fuzzy Logic, and Fuzzy Systems: Selected Papers*, vol. 6 (World Scientific, Singapore, 1996)

[ZZ80] H. Zimmermann, P. Zysno, Latent connectives in human decision making. Fuzzy Sets Syst. **4**, 37–51 (1980)

CPSIA information can be obtained at www.ICGtesting.com
Printed in the USA
LVOW10*1441030314

375821LV00013B/166/P

9 783319 042251